Lecture Notes in Physics

Editorial Board

H. Araki
Research Institute for Mathematical Sciences
Kyoto University, Kitashirakawa
Sakyo-ku, Kyoto 606, Japan

E. Brézin
Ecole Normale Supérieure, Département de Physique
24, rue Lhomond, F-75231 Paris Cedex 05, France

J. Ehlers
Max-Planck-Institut für Physik und Astrophysik, Institut für Astrophysik
Karl-Schwarzschild-Strasse 1, W-8046 Garching, FRG

U. Frisch
Observatoire de Nice
B. P. 229, F-06304 Nice Cedex 4, France

K. Hepp
Institut für Theoretische Physik, ETH
Hönggerberg, CH-8093 Zürich, Switzerland

R. L. Jaffe
Massachusetts Institute of Technology, Department of Physics
Center for Theoretical Physics
Cambridge, MA 02139, USA

R. Kippenhahn
Rautenbreite 2, W-3400 Göttingen, FRG

H. A. Weidenmüller
Max-Planck-Institut für Kernphysik
Postfach 10 39 80, W-6900 Heidelberg, FRG

J. Wess
Lehrstuhl für Theoretische Physik
Theresienstrasse 37, W-8000 München 2, FRG

J. Zittartz
Institut für Theoretische Physik, Universität Köln
Zülpicher Strasse 77, W-5000 Köln 41, FRG

Managing Editor

W. Beiglböck
Assisted by Mrs. Sabine Landgraf
c/o Springer-Verlag, Physics Editorial Department V
Tiergartenstrasse 17, W-6900 Heidelberg, FRG

The Editorial Policy for Proceedings

The series Lecture Notes in Physics reports new developments in physical research and teaching – quickly, informally, and at a high level. The proceedings to be considered for publication in this series should be limited to only a few areas of research, and these should be closely related to each other. The contributions should be of a high standard and should avoid lengthy redraftings of papers already published or about to be published elsewhere. As a whole, the proceedings should aim for a balanced presentation of the theme of the conference including a description of the techniques used and enough motivation for a broad readership. It should not be assumed that the published proceedings must reflect the conference in its entirety. (A listing or abstracts of papers presented at the meeting but not included in the proceedings could be added as an appendix.)
When applying for publication in the series Lecture Notes in Physics the volume's editor(s) should submit sufficient material to enable the series editors and their referees to make a fairly accurate evaluation (e.g. a complete list of speakers and titles of papers to be presented and abstracts). If, based on this information, the proceedings are (tentatively) accepted, the volume's editor(s), whose name(s) will appear on the title pages, should select the papers suitable for publication and have them refereed (as for a journal) when appropriate. As a rule discussions will not be accepted. The series editors and Springer-Verlag will normally not interfere with the detailed editing except in fairly obvious cases or on technical matters.
Final acceptance is expressed by the series editor in charge, in consultation with Springer-Verlag only after receiving the complete manuscript. It might help to send a copy of the authors' manuscripts in advance to the editor in charge to discuss possible revisions with him. As a general rule, the series editor will confirm his tentative acceptance if the final manuscript corresponds to the original concept discussed, if the quality of the contribution meets the requirements of the series, and if the final size of the manuscript does not greatly exceed the number of pages originally agreed upon. The manuscript should be forwarded to Springer-Verlag shortly after the meeting. In cases of extreme delay (more than six months after the conference) the series editors will check once more the timeliness of the papers. Therefore, the volume's editor(s) should establish strict deadlines, or collect the articles during the conference and have them revised on the spot. If a delay is unavoidable, one should encourage the authors to update their contributions if appropriate. The editors of proceedings are strongly advised to inform contributors about these points at an early stage.
The final manuscript should contain a table of contents and an informative introduction accessible also to readers not particularly familiar with the topic of the conference. The contributions should be in English. The volume's editor(s) should check the contributions for the correct use of language. At Springer-Verlag only the prefaces will be checked by a copy-editor for language and style. Grave linguistic or technical shortcomings may lead to the rejection of contributions by the series editors. A conference report should not exceed a total of 500 pages. Keeping the size within this bound should be achieved by a stricter selection of articles and not by imposing an upper limit to the length of the individual papers. Editors receive jointly 30 complimentary copies of their book. They are entitled to purchase further copies of their book at a reduced rate. As a rule no reprints of individual contributions can be supplied. No royalty is paid on Lecture Notes in Physics volumes. Commitment to publish is made by letter of interest rather than by signing a formal contract. Springer-Verlag secures the copyright for each volume.

The Production Process

The books are hardbound, and the publisher will select quality paper appropriate to the needs of the author(s). Publication time is about ten weeks. More than twenty years of experience guarantee authors the best possible service. To reach the goal of rapid publication at a low price the technique of photographic reproduction from a camera-ready manuscript was chosen. This process shifts the main responsibility for the technical quality considerably from the publisher to the authors. We therefore urge all authors and editors of proceedings to observe very carefully the essentials for the preparation of camera-ready manuscripts, which we will supply on request. This applies especially to the quality of figures and halftones submitted for publication. In addition, it might be useful to look at some of the volumes already published. As a special service, we offer free of charge LATEX and TEX macro packages to format the text according to Springer-Verlag's quality requirements. We strongly recommend that you make use of this offer, since the result will be a book of considerably improved technical quality. To avoid mistakes and time-consuming correspondence during the production period the conference editors should request special instructions from the publisher well before the beginning of the conference. Manuscripts not meeting the technical standard of the series will have to be returned for improvement.

For further information please contact Springer-Verlag, Physics Editorial Department V, Tiergartenstrasse 17, W-6900 Heidelberg, FRG

K. H. Ploog   L. Tapfer   (Eds.)

Physics and Technology of Semiconductor Quantum Devices

Proceedings of the International School
Held in Mesagne (Brindisi), Italy
21-26 September 1992

Springer-Verlag
Berlin Heidelberg New York
London Paris Tokyo
Hong Kong Barcelona
Budapest

Editors

Klaus H. Ploog
Paul-Drude-Institut für Festkörperelektronik
Hausvogteiplatz 5-7, D-10117 Berlin, Germany

Leander Tapfer
Centro Nazionale per la Ricerca e lo Sviluppo dei Materiali
S.S.7 per Mesagne Km. 7, 300, I-72100 Brindisi, Italy

ISBN 3-540-56989-8 Springer-Verlag Berlin Heidelberg New York
ISBN 0-387-56989-8 Springer-Verlag New York Berlin Heidelberg

This work is subject to copyright. All rights are reserved, whether the whole or part of the material is concerned, specifically the rights of translation, reprinting, re-use of illustrations, recitation, broadcasting, reproduction on microfilms or in any other way, and storage in data banks. Duplication of this publication or parts thereof is permitted only under the provisions of the German Copyright Law of September 9, 1965, in its current version, and permission for use must always be obtained from Springer-Verlag. Violations are liable for prosecution under the German Copyright Law.

© Springer-Verlag Berlin Heidelberg 1993
Printed in Germany

Typesetting: Camera ready by author/editor
58/3140-543210 - Printed on acid-free paper

Preface

The lecture papers contained in this volume were presented at the International School on Physics and Technology of Semiconductor Quantum Devices, which was held in September 21-26, 1992 in Mesagne (Brindisi), Italy.

The School was organized within the Training Program on Materials Science of the Centro Nazionale per la Ricerca e lo Sviluppo dei Materiali (CNRSM).

It was organized for the personnel of the CNRSM and open also to young scientists from all over Europe with a few years of experience in the physics and technology of semiconductors.

The purpose of the School was to present overviews of the basic concepts, current research and research developments of semiconductor quantum devices to graduate and postgraduate students.

The chapters of this book are written by scientists well known for their contributions to the development of their specific research field. The book covers a wide range of the scientific research in the field of semiconductor quantum devices and focuses on links between physics and technology. Particular emphasis is given to the technology, fabrication and physics of electronic and opto-electronic devices.

The School was sponsored by: Agenzia per la Promozione dello Sviluppo del Mezzogiorno (Roma), IBM-SEMEA (Milano), and Philips Spa (Monza). The financial support of these institutions is gratefully acknowledged.

We wish to thank all lecturers for their excellent work and for their willingness to cooperate within a very limited timetable. We are particularly indebted to Prof. Paolo Cavaliere, President of the CNRSM, and to Prof. Lorenzo Vasanelli, Director of the Training Program. Our special thanks go to the School Secretary, Dr. Rosanna Cramarossa, for her never-tiring organizational activity that made our School a real success.

Klaus H. Ploog
Leander Tapfer

Contents

Klaus H. Ploog
Epitaxial Growth of Nanostructured III-V Semiconductors............1

Sybrand Radelaar
*Technology and Fabrication of Quantum Devices:
Submicron Lithography and Etching Techniques*............27

Roberto Cingolani
Band Gap Engineering in Low-Dimensional Semiconductors............53

Joachim Wagner
*Electronic and Optical Properties of Low-Dimensional
Semiconductor Structures*............71

Nicolas J. Pulsford
*Electronic Structure and Electrical Characterisation of
Semiconductor Heterostructures*............97

Emmanuel Rosencher
New Optoelectronic Devices............127

Michael J. Kearney
High Speed Devices: Properties and Applications............151

Klaus H. Ploog and Richard Nötzel
*New Concepts to Fabricate Semiconductor Quantum Wire and
Quantum Dot Structures*............199

Epitaxial Growth of Nanostructured III-V Semiconductors

Klaus H. Ploog

Paul-Drude-Institut für Festkörperelektronik
O - 1086 Berlin, Federal Republic of Germany

1 Introduction

At about the same time as Esaki and Tsu [1] predicted the intriguing properties of semiconductor superlattices, a new thin film growth technique, called molecular beam epitaxy (MBE), was developed [2] to grow ultrathin layers and periodic multilayer structures of III-V semiconductors in a reproducible manner. Today, several subcategories of beam epitaxy and vapor phase epitaxy (VPE) have been developed to fabricate routinely artificially layered semiconductors of high structural perfection for fundamental research and for device applications. In Fig. 1 we show that in ultrathin strained InAs layers in a GaAs matrix the modulation in chemical composition is abrupt within atomic dimensions. The accurate positioning of heterojunctions provided by advanced thin film growth techniques defines the desired potential differences in semiconductor microstructures locally on nanometer scale. In this paper we first describe the fundamentals of several epitaxial methods for growth of III-V compounds, including metalorganic vapor phase epitaxy (MO VPE), conventional elemantal source molecular beam epitaxy (ES MBE), and the various subcategories of gas-source molecular beam epitaxy (GS MBE). We then briefly discuss methods for additional lateral patterning of quantum well structures leading to a new class of exotic microstructures such as quantum wires and quantum dots.

2 Fundamentals of Beam Epitaxy and of Metalorganic Vapor Phase Epitaxy

The particular merits of the crystal growth techniques discussed here are that ultrathin layers can be grown with precise control over thickness, composition and doping level. Molecular beam epitaxy using elemental sources is in principle a rather elementary synthetic process, because only surface phenomena are involved in the crystal growth, no foreign atoms are present at the gas-solid interface, and by-products are not formed. Our understanding of the basic growth mechanisms invol-

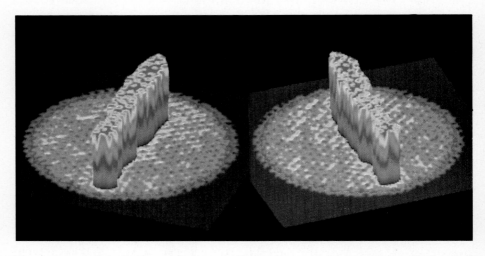

Fig. 1 Color-coded 3D representation of a lattice image of 2 ML InAs in a GaAs matrix grown by ES MBE. The perspective shows the structural equivalence of both interfaces.

ved in elemantal source MBE (ES MBE) is very advanced [3,4] and techniques to monitor and control atomic layer deposition are readily available (reflection high energy electron diffraction (RHEED) and reflectance difference spectroscopy (RDS)). On the contrary, the basic processes occurring in MO VPE, that involve complex combinations of gas phase and surface phenomena, are less well understood, and monitoring techniques are only currently being developed [5]. MO VPE has a decided advantage in the growth of phosphorus containing III-V compounds, and it is widely used to fabricate $Ga_xIn_{1-x}P_yAs_{1-y}/InP$ heterostructures and quantum wells. However, since compounds are used as precursors in MO VPE, a great effort has to be expended in order to achieve the required levels of high purity and safety/convenience of the growth process [6].

The use of any gaseous sources, whether the group III or group V in the form of alkyls and/or hydrides, in conventional MBE systems has introduced a further dimension to molecular beam epitaxy [7,8]. The potential versatility of source materials can be combined with the extremely sharp interfaces and the availability of monitoring techniques associated with molecular beam epitaxy. Depending on the various combinations of sources, the subcategories of beam epitaxy can be classified as illustrated in Fig. 2. Although different classification schemes and differences in the terminology exist, we use the convention that the term gas-source MBE (GS MBE) denotes the use of any gaseous sources, whether group III and/or group V, to generate the beams for epitaxy. The distinction between beam epitaxy and vapor phase epitaxy (VPE) is then made primarily about the pressure regimes involved in each method. Reasonable criteria for beam epitaxy are that the pressure be low enough that the transport of atoms and molecules is by molecular flow and that the minimum free path to be no less than the source-to-substrate distance. The upper

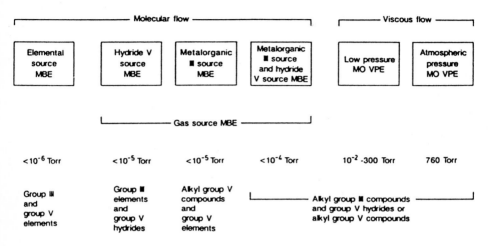

Fig. 2 Classification of vapor phase epitaxy and the various subcategories of beam epitaxy according to the pressure regimes involved in each method.

pressure limit for molecular flow in a typical MBE system is found to be about 10^{-3} Torr, in agreement with simple criteria provided earlier by Dushman [9] for determining the upper pressure limit for molecular flow and the lower pressure limit for viscous flow of air. Although the precise pressure limits are not sharply defined, under molecular-flow conditions there is clearly no boundary layer at the substrate-gas interface. The loss of this boundary layer may imply, however, that useful boundary layer reactions which limit carbon incorporation from the metal alkyls may also be lost. Although there are significant differences in the complexity of gas handling systems for toxic and hazardous chemicals and in the growth mechanisms between the various beam techniques, several GS MBE subcategories now compare favorably with ES MBE particularly for those III-V semiconductors containing phosphorus [10].

3 Growth Techniques
3.1 Elemental Source Conventional Molecular Beam Epitaxy

The MBE technique allows lattice-plane by lattice-plane deposition of custom-designed microstructures in a two-dimensional (2D) growth process [11,12]. Crystalline materials in alternating layers of arbitrary composition and only a few atomic layers thick can thus be fabricated in a reproducible manner [13]. Of particular importance is that the monocrystalline pattern of the lattice unit in successive layers continues without disruption or distortion across the interfaces between the layers. The basic process for ES MBE growth of III-V semiconductors consists of a co-evaporation of the constituent elements (Al, Ga, In, P, As, Sb) of the epitaxial layer and of dopants (mainly Be for p-type and Si for n-type doping) on to a heated crystalline substrate where they react chemically under ultra-high vacuum (UHV) conditions (Fig. 3). The composition of the layer and its doping level depend mainly on the relative arrival rates of the constituent elements which in turn depend on the evaporation rates of the respective sources. Accurately adjusted temperatures (to within $+0.1^{\circ}C$ at $1000^{\circ}C$) have thus a direct and controllable effect upon the growth process.

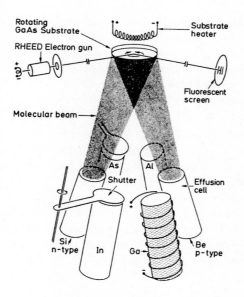

Fig.3   Schematic illustration of evaporation and deposition process in MBE of III-V compounds in a UHV system equipped with resistively heated effusion cells with shutters, the substrate which is rotating for better homogeneity, RHEED to determine the structure of the growing surface, and LN_2 cryopanels (not shown) encircling growth area and effusion cells.

The group III elements are always supplied as monomers by evaporation from the respective liquid element, and they have a unity sticking coefficient over most of the substrate temperature range used for film growth (e.g. 500-630°C for GaAs) [14]. The group V elements on the other hand are supplied as tetramers (P_4, As_4, Sb_4) by sublimation from the respective solid element or as dimers (P_2, As_2, Sb_2) by dissociating the tetrameric molecules in a cracker a two-zone furnace. The film growth rate is typically 0.5-1.0 μm/h. It is chosen low enough that dissociation and migration of the impinging species on the growing surface to the appropriate lattice sites are ensured without incorporating crystalline defects. Simple mechanical shutters in front of the evaporation sources are used to interrupt the beam fluxes in order to start and stop deposition and doping. Due to the slow growth rate of about one lattice plane per second, changes in composition and doping can thus be abrupt on an atomic scale. This independent and accurate control of the individual beam sources allows the precise fabrication of artificially layered semiconductor structures on an atomic scale, as revealed for example by transmission electron microscopy.

The stoichiometry of most III-V semiconductors during MBE growth is self-regulating as long as excess group V molecules are impinging on the growing surface. The excess group V species do not stick on the heated substrate surface, and the growth rate of the films is essentially determined by the arrival rates of the group III elements [14]. A good control of ternary III-III-V alloys can thus be achieved by supplying excess group V species and adjusting the flux densities of the impinging group III beams, as long as the substrate temperature is kept below the congruent evaporation limit of the less stable of the constituent binary III-V compounds (i.e. GaAs in the case of $Al_xGa_{1-x}As$). However, at higher growth temperature preferential desorption of the more volatile group III element (i.e. Ga from $Al_xGa_{1-x}As$) occurs so that the final film composition is not only determined by the added flux ratios but also by the differences in the desorption rates. To a first approximation we can estimate the loss rate of the group III elements from their vapor pressure data. The surface of alloys grown at high temperatures will thus be enriched in the less volatile group III elements. As a consequence, we observe a significant loss of In in $Ga_xIn_{1-x}As$ layers grown above 550°C and a loss of Ga in $Al_xGa_{1-x}As$ layers grown above 650°C. The growth of ternary III-V-V alloys, like GaP_yAs_{1-y}, by MBE is more complicated, as even at moderate substrate temperatures the relative amounts of the group V elements incorporated into the growing layer are not simply proportional to their relative arrival rates. The factors controlling this incorporation behavior are at present not understood [14], nor for the other beam epitaxy methods depicted in Fig. 2. It is therefore difficult to obtain a reproducible composition control during

ES MBE growth of ternary III-V-V alloys, especially when the tetrameric group V species are used. In ternary III-III-V alloys, on the other hand, the simplicity of the MBE process allows compositional control from x=0 to x=1 in $Al_xGa_{1-x}As$, $Ga_xIn_{1-x}As$, etc., with a precision of + 0.001 and doping control, both p- and n-type, from the 10^{13} cm^{-3} to the 10^{19} cm^{-3} range with a precision of a few percent. The accuracy is largely determined by the care with which the growth rate and doping level were precisely calibrated in test layers.

The purity of most III-V semiconductors grown by ES MBE is limited by background impurities originating from the UHV system and from the source materials. Not intentionally doped GaAs layers with residual impurity concentrations in the low 10^{13} cm^{-3} range are now routinely achieved. The most common dopants used during MBE growth are Be for p-type and Si for n-type doping. The group II element Be behaves as an almost ideal shallow acceptor in many MBE grown III-V semiconductors [15]. Each incident Be atom produces one ionizes impurity species, providing an acceptor level 29 meV above the valence band edge in GaAs. At doping concentrations above 3×10^{19} cm^{-3}, however, the GaAs surface morphology and electronic properties degrade, and the diffusion of Be is enhanced [16]. Recently, carbon which has always been an unintentional background impurity of MBE and MO VPE grown III-V semiconductors has been recognized as an attractive p-type dopant substitute for Be because of its high solid solubility ($>10^{20}$cm^{-3}) and its extremely low diffusion coefficient in GaAs and $Al_xGa_{1-x}As$ [17]. Intentional carbon incorporation during MO MBE and MO VPE has been achieved through incomplete cracking of the metal-carbon bonds of metalorganics and during ES MBE through carbon evaporation from a resistively heated solid graphite source. In particular, the device characteristics of GaAs/$Al_xGa_{1-x}As$ heterojunction bipolar transistors have been greatly improved using a heavily carbon doped p-type base.

The group IV element Si is primarily incorporated on group III sites during MBE growth under As-stabilized conditions, yielding n-type material of fairly low compensation. The observed doping level is simply proportional to the dopant arrival rate, provided care is taken to reduce the H_2O and CO background level during growth [18]. The upper limit of $n = 1 \times 10^{19}$ cm^{-3} for the free-electron concentration in GaAs is given by the formation of [Si-X] complexes [19]. The possibility of Si migration during MBE growth of $Al_xGa_{1-x}As$ layers at high substrate temperatures and/or with high donor concentrations results from a concentration-dependent diffusion process which is enhanced at high substrate temperatures [20]. The distribution of the dopant impurities can be confined to regions as narrow as a few lattice planes in the direction of growth. This monolayer- or delta-doping is obtained by interruption of the growth, and it generates V-shaped potential wells in the host material. The electronic properties of delta-doped semiconductors are very different from homogeneously doped semiconductors, and this doping concept has thus become important for fundamental research as well as for device applications [21]. It is finally noteworthy that the incorporation of Si atoms on either Ga or As sites during MBE growth depends strongly on the orientation of the GaAs substrate [22]. In GaAs deposited on (111)A, (211)A and (311)A orientations the Si atoms predominantly occupy As sites and act as acceptors, whereas they occupy Ga sites and act as donors on (100), (111)B, (211)B, (311)B, (511)A, (511)B and higher-index orientations. However, very recent studies have shown [23] that most of the non-(100)-oriented GaAs surfaces are no longer flat but form macrosteps/facets under typical MBE growth conditions (see Sect. 4 for details).

The advanced MBE systems mostly consist of three basic UHV building blocks, i.e. the growth chamber, the sample preparation chamber, and the load-lock chamber, which are separately pumped and interconnected via large-diameter channels and isolation valves [24]. High-quality layered semiconductor structures require a background vacuum in the low 10^{-11} Torr range to avoid incorporation of background impurities into the growing layers. Therefore, extensive LN_2 cryoshrouds are used around the substrate to achieve locally much lower background pressures of condensible species. The starting materials for the growth of III-V semiconductors are evaporated in resistively-heated effusion cells made of pyrolytic BN which operate at temperatures up to 1400°C. Most of the functions important for the MBE growth process are controlled by a computer.

Molecular beam epitaxy of III-V semiconductors is mostly performed on (001) oriented substrate slices about 300-500 μm thick. The preparation of the growth face of the substrate from the polishing stage to the *in-situ* cleaning stage in the UHV system is of crucial importance for epitaxial growth of ultrathin layers and heterostructures with high purity and crystal perfection and with accurately controlled interfaces on an atomic scale. The substrate surface should be free of crystallographic defects and clean on an atomic scale. Various cleaning methods have been described for GaAs and InP [2,12,24] which are the most important substrate materials for deposition of III-V semiconductors. The first step always involves chemical cleaning and etching, which leaves the surface covered with some kind of a protective oxide. After insertion in the MBE system this oxide is removed by heating under UHV conditions carried out in a beam of the group V element.

The most important method to monitor *in-situ* surface crystallography and growth kinetics during MBE is reflection high-energy electron diffraction (RHEED) operated at 10-50 keV in the small glancing angle reflection mode (Fig. 4) [25]. The diffraction pattern on the fluorescent screen, mostly taken in the [110] and [1̄10] azimuths of (001) oriented substrates, contains information from the topmost layers of the deposited material, and it can thus be related to the topography and structure of the growing surface. The diffraction spots are elongated to charac-

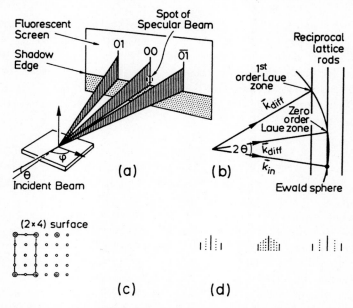

Fig. 4 Geometric arrangement of RHEED with grazing-angle incidence used as *in-situ* analytical tool during MBE(a), Ewald construction to interpret the diffraction pattern (b), surface unit cell ("mesh")(c), and diffraction patterns (d) of the most importants As-stabilized (2 x 4) reconstruction during epitaxy.

teristic streaks normal to the shadow edge. Additional features in the RHEED pattern at fractional intervals between the bulk diffraction streaks manifest the existence of a specific surface reconstruction, which are correlated to the surface stoichiometry and thus directly to the MBE growth conditions (substrate temperature, molecular beam flux ratio etc.) [2,3,4].

Another characteristic feature of the RHEED pattern is the existence of pronounced periodic intensity oscillations of the specularly reflected and of several diffracted beams during MBE growth [26,27]. The period of these oscillations corresponds exactly to the time required to deposit a lattice plane of GaAs (or AlAs, $Al_xGa_{1-x}As$, etc.) on the (001) surface (Fig. 5). This peculiarity is explained by the assumption that the amplitude of the intensity oscillation reaches its maximum when the

lattice plane is just completed (maximum reflectivity).The formation of
the following lattice plane starts with statistically distributed 2D
islands having the height of one GaAs lattice plane [0.28 nm for the
(001) plane]. The intensity of the reflected (or diffracted) electron
beam decreases with increasing size of the islands. The minimum
reflectivity occurs at half-layer coverage ($\Theta = 0.5$). When the coverage
is further increased the islands coalesce more and more, and the reflec-

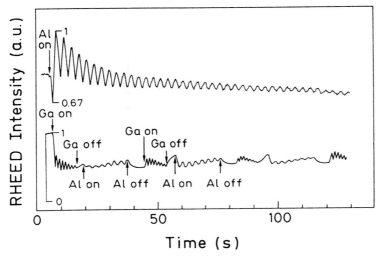

Fig. 5 Periodic intensity oscillation of the specularly-reflected electron beam in the RHEED pattern as a function of time during growth of GaAs/AlAs layers.

tivity reaches a maximum again at $\Theta = 1$. The observed intensity oscillations in the RHEED pattern provide direct evidence that MBE growth occurs predominantly in a 2D layer-by-layer growth mode. This method is
now widely used to calibrate and to monitor absolute growth rates in
real time with monolayer resolution. However, it is important to note
that the interpretation of RHEED intensity data may be erroneous because
in general both elastically and diffusely scattered electrons contribute
to the recorded intensity. Diffraction conditions where elastic scattering dominates the intensity of the specularly reflected beam are most
easily obtained under off-azimuth conditions. The other important aspect
is the damping of the RHEED intensity oscillations observed during
growth and the characteristic recovery once growth has been interrupted.
The damping has been ascribed [28] to an increase of the surface step
density, i.e. an increasing number of individual surface domains probed
by the electron beam are no longer in phase. The recovery of the RHEED
intensity following interruption of growth has been identified with an
expansion of the mean terrace width of the surface and hence a reduction

of the surface step density, i.e. a few large domains on the growing surface were formed.

Recently, an *in-situ* optical technique called reflectance difference spectroscopy (RDS) has been introduced [29] to monitor the surface chemistry during MBE growth of GaAs and $Al_xGa_{1-x}As$ layers and heterostructures. This technique is primarily sensitive to the anisotropy of the growing surface induced by Ga-Ga dimers and hence to the surface stoichiometry. In combination with RHEED a careful analysis of RDS data provides detailed insights into the MBE growth mechanisms of III-V compounds.

3.2 Metalorganic Vapor Phase Epitaxy

The particular epitaxy process that involves the pyrolysis of vapor-phase mixtures of metalorganic group III compounds and group V hydrides with hydrogen carrier gas was pioneered by Manasevit [30]. In general the trimethyl- or triethyl-group III compounds (TMAl, TMGa, TMIn, TEAl, TEGa, TEIn, etc.) are used in combination with the group V hydrides PH_3 and AsH_3. Pyrolysis (substrate) temperatures typically in the range 600 to 800°C are employed [5].

The detailed chemical reaction paths leading to the epitaxial growth of materials by MO VPE are still not well known and the physical kinetics of deposition have not been studied in great detail. Nevertheless, MO VPE has become a versatile technique to fabricate a variety of III-V heterostructures for device application and for fundamental studies. The growth of GaAs by MO VPE can be described by the following overall reaction:

$$(CH_3)_3Ga + AsH_3 = GaAs + 3CH_4.$$

The most probable intermediate steps involve the thermal and/or surface catalyzed decomposition of the starting reactants followed by recombination of the Ga and As on the substrate surface. As illustrated in Fig. 6, the gas flow dynamics in a reactor at or near atmospheric pressure leads to the formation of a boundary layer at the stationary substrate surface. In the gas-flow direction the thickness of this boundary layer increases. The reduced gas velocity in this region results in a stagnant layer of gases which controls the transport of gaseous species to the substrate surface. The reactants are consumed when they reach the surface. A steady-state gradient is thus established at any point along the surface which drives the diffusion of the reactants. The boundary layer is also characterized by a strong temperature gradient through

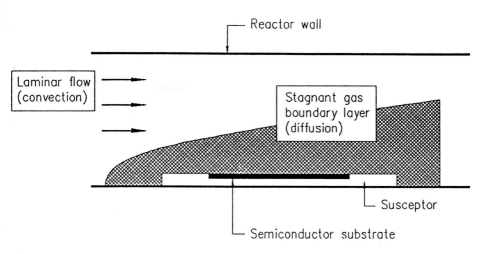

Fig. 6 Schematic illustration of the gas flow in a horizontal VPE reactor.

which the reactants diffuse and where they can decompose to release the atomic species needed for film growth. If the diffusion rate through the boundary layer exceeds the decomposition rate of the reactants then catalytic effects at the surface may become important in the film growth. Problems may arise from reactions between the group III metalorganics and group V hydrides in the cold gas during transport to the hot growth zone. The formation of nonvolatile adducts between TEIn and PH_3, e.g., leads to a depletion of reactants in the carrier gas and hence a lowered growth rate or shifted alloy compostion. At relatively low concentration ($>10^{-4}$ mole fraction) and high gas velocities in the reactor, the reaction may only progress a few percent toward completion before the reactants reach the pyrolysis zone.

The growth of III-V semiconductors by MO VPE requires also an excess of the group V reactants, because of the volatility of the group V elements. Under these conditions the growh rate is proportional to the flow of the group III reactants. The purity of III-V semiconductors grown by MO VPE depends not only on the purity of the starting materials but also critically on the ratio of the group III and group V fluxes on the growth temperature and on the reactor pressure. Using optimized growth conditions total impurity concentrations in the low 10^{14} cm^{-3} range have been obtained for GaAs layers. The situation is much more difficult for ternary $Al_xGa_{1-x}As$ layers because of the extreme reactivity of aluminium. Great care must be taken that the reactor is leak-tight and the growth temperature has to be increased above 700°C. For the growth of ternary III-III-V semiconductors metalorganic group III precursors should be selected having similar masses and therefore similar diffusion coefficients in order to minimize alloy composition gradients [31].

Doping in MO VPE is accomplished by introducing the appropriate gaseous reactants into the gas flow [5]. These reactants diffuse through the boundary layer in the same way as the main constituent reactants. The incorporation of the dopants is then controlled either by surface reactions (adsorption vs desorption or surface catalysis) or the thermochemistry of the reactants. For p-type doping metalorganic compounds of Zn or Mg are used, while the hydrides of Se or Si are used as n-type dopants. In the case of these metalorganic dopants, surface kinetics control their incorporation as the compounds decompose readily at the growth temperature. As for the n-type dopant hydrides, the thermochemistry of the hydrides controls the doping efficiency.

The growth of III-V multilayer structures in MO VPE is accomplished by changing the gas composition in the reactor. The rate at which such a change occurs depends on the total flow and on the geometry of the reac-

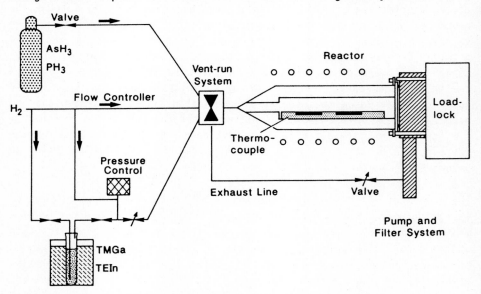

Fig. 7 Schematic sketch of the essential components of a MO VPE system.

tor. At high flow rates, this exchange can be accomplished rapidly enough to form atomically abrupt heterojunctions. The time constant for the exchange is controlled by the total flow rate. As a consequence, abrupt GaAs/Al$_x$Ga$_{1-x}$As interfaces have been grown both by interrupting the growth and flushing the reactor and by continuous growth at high flow rates and low growth rates. Sophisticated gas manifolds of today make feasible fast switching of the reactants and hence abrupt changes in film composition perpendicular to the substrate surface. Operation of MO VPE at reduced pressure (10 to 300 Torr) results in an improved homogeneity of composition and thickness of the deposited layers [32,33]. In low-pressure (LP) MO VPE the local variation of the gas-phase composition is reduced by the enhanced diffusivity. In addition, with respect to unwanted reactions between precursors, low-pressure operation effectively reduces the contact time and minimizes the extents of those reactions. The method has been particularly successful for growth of $Ga_xIn_{1-x}P_yAs_{1-y}$/InP heterostructures and quantum wells (33). However, the group V hydride consumption rises dramatically in low-pressure MO VPE, as it is the pressure of the group V species that must be held constant over the substrate surface.

MO VPE systems may be grouped into four major components: [5] (i) the gas handling system including the source metalorganics and hydrides and all of the valves plumbing and instruments necessary to control the gas flows and mixtures; (ii) the reactor chamber in which the pyrolytic reaction and the deposition occurs; (iii) the heating system to obtain the pyrolysis temperature; (iv) the exhaust or low-pressure pumping system. The quartz reactors used in MO VPE typically consist of an outer and an inner liner tube (Fig. 6) to facilitate removal of large amounts of arsenic wall deposits. The substrate wafer is placed on a graphite susceptor which is externally heated either by radio-frequency (RF) induction or by radiation from quartz halogen lamps. The first heating method is less suitable for LP MO VPE because of the uncontrolled formation of an RF plasma at lower total pressures. The gas mixture near the substrate surface is thus heated to high temperatures while the walls of the reactor remain relatively cool ("cold wall system"). This temperature distribution leads to deposition of the semiconductor layer on the substrate and not much loss of reactants to the surface of the reactor chamber. At present both horizontal and vertical-flow reactors are employed which have a fundamental feature in common. They exhibit distinct profiles in flow velocity, concentration of the reactants and temperature which strongly depend on the applied reactor geometry. The reaction conditions over the entire substrate area should essentially be constant in order to achieve homogeneous deposition. This condition implies that the composition of the gas phase should not significantly change over the substrate area, especially for the growth of ternary and quaternary semiconductors. The important consequence is that the largest amount of the starting materials leaves the reaction zone unused.

The accurate and reproducible control of the flows of the gaseous starting materials is essential for the deposition of heterostructures, in particular when lattice matching is required. The gas handling system

(Fig. 7) has thus to deliver precisely metered amounts of uncontaminated reactants without any transients due to pressure or flow charges. In general purified hydrogen is used to transport the low vapor pressure metalorganics from their stainless steel container, held in temperature-controlled baths, into the reactor. The flux of hydrogen is controlled upstream of this container by an electronic mass flow controller (MFC). To keep the pressure inside the container constant, irrespective of the total pressure in the reactor, a pressure control circuit consisting of a pressure transducer (mostly a capacitance manometer) and a servovalve is added. The flow of the reactants having higher vapor pressure (PH_3, As_3) or hydrogen-diluted dopants are directly controlled by the MFCs. The gas handling systems are typically assembled from stainless steel tubing and precision valves. The MFCs as well as other critical valves are automated and interfaced to a computer.

Today's MFCs applied in MO VPE exhibit a relative accuracy of 1 % at full scale. However, their limited dynamic range reduces the accuracy in growing heterostructures with graded composition at the interfaces. On the other hand, the pressure balanced run-vent lines in advanced MO VPE systems allow fast switching of the reactants into the reactor without pressure or flux transients. With the successful growth of $(GaAs)_m/(AlAs)_n$ ultrathin-öayer superlattices. Ishibashi et al. [34] have demonstrated that this technique reproducibly ensures sharp interfaces by abruptly changing from one conposition to the other without growth interruption. Also the first GaAs quantum well lasers operating continuosly at room temperature habe been produced from MO VPE grown wafers [35].

In low-pressure MO VPE, often used for improved homogeneity over large substrate diameters as well as for reduced consumption of toxic reactants [33], a two-stage rotary pump is normally employed. A thermal cracker combined with a cold trap is added between reactor and pump in order to protect the user (and the environment) from toxic gases which may also accumulate and react in the pumping liquid and reduce the pumping speed. The exhaust gasses of the pump are finally passed through a scrubbing system before entering the laboratory exhauster. As yet little effort has been devoted to the *in-situ* monitoring of the chemistry and the growth process in MO VPE. Raman scattering [36] and surface science probes [37] have been used to study the chemistry, and the surface photoabsorption method [38] has been employed for *in-situ* monitoring of epitaxial growth.

3.3 Gas-Source Molecular Beam Epitaxy

In the various subcategories of gas-source molecular beam epitaxy (GS MBE) the elemental sources are replaced by volatile metalorganic compounds and/or hydrides to transport the constitutents to the growing gas-solid interface. In general the same reactants as in MO VPE are used. The replacement of the elemental sources in conventional MBE by gaseous source materials was initiated by the search for a long-lasting arsenic source and for a reproducible composition control during growth of ternary III-V-V alloys based on phosphorus and arsenic [39]. Thermal cracking of the hydrides PH_3 and AsH_3 at temperatures around 900°C produces the dimers P_2 and As_2 as well as hydrogen. This hydride-source MBE (HS MBE) [7], where only the group V elements are replaced by their respective hydrides, allows epitaxial growth of $Ga_{0.47}In_{0.53}As/InP$ heterostructures and superlattices of high quality and of $Ga_xIn_{1-x}P_yAs_{1-y}/InP$ heterostructures with reasonable composition control. An extension of this concept was then made by replacement of the group III elements by metalorganic compounds [40]. In this metalorganic MBE (MO MBE) and chemical beam epitaxy (CBE) [8] the metalorganic flows mixed with hydrogen are in some cases combined outside the UHV growth chamber to form a single beam impinging onto the heated substrate for good compositional uniformity across the substrate area. On the heated substrate surface thermal pyrolysis of the metalorganic compounds takes place and in an excess group V beam the III-V semiconductor is formed. Also these techniques allow reproducible growth of P-containing III-V semiconductor heterostructures of high quality.

The concentration of toxic gases in the exhaust is much lower in GS MBE than in MO VPE, since the highly toxic group V hydrides are effectively cracked in the hot zone of the cell and their consumption is low.

In UHV systems used for GS MBE the effusion cells are replaced by metalorganic entry tubes and/or by group V gas-source cracker (Fig. 8).To provide for adequate pumping of the large amounts of hydrogen, a throughput turbomolecular or diffusion pump of sufficient pumping speed has to be attached to the UHV system. One of the most important components of GS MBE systems is the gas handling system that provides means of regulating the flow rates of the starting compounds into the growth chamber. Precision mass flow controllers with linear response are used to control directly the flow of the pure gaseous group V hydrides into the thermal cracker. The metalorganics are transported by hydrogen carrier gas of fixed pressure bubbling through the (liquid) metalorganics kept at accurately controlled temperatures. Mass flow control is then of the combined gas stream. Another approch [8] avoids hydrogen carrier gas for the metalorganics. Instead, pressure transducers and direct evaporation from the temperature-controlled containers are employed. This method keeps the hydrogen load in the growth chamber and hence the total pressure as low as possible. For doping the elements Be an Si evaporated in conventional effusion cells are used also in gas-source molecular beam epitaxy [7,8]. It is obvious that there are significant differences in growth mechanisms between the various subcategories of MBE, due to the use of metalorganic species [41]. In solid-source and hydride-source MBE there is no interaction in the beams, and the growth rate depends only on the arrival rate of the (elemental) group III species [14]. Since these have unity sticking coefficients up to a certain critical temperature, the growth rate is relatively independent of substrate temperature. In MO MBE and CBE, on the other hand, the substrate temperature adopts two functions. Firstly. it has to decompose the metalorga-

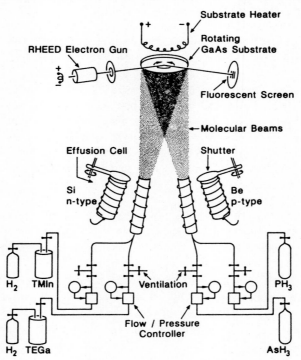

Fig. 8 Schematic illustration of a gas-source MBE system using volatile metalorganic compounds and hydrides as well as the elemental dopants Si and Be.

nics and then, secondly, to impart energy to ensure sufficient atom mobility for epitaxial growth. The growth rate now depends on substrate temperature. At low temperatures, the growth rate is limited by the efficiency of the pyrolysis of the metalorganics. Then a plateau range exists where this pyrolysis occurs at a constant rate. Unlike in ES MBE, the mobile surface species is a metal alkyl with at least one alkyl group removed. Recent studies of the surface chemistry confirm that the cleavage of the metal-alkyl bonds occurs sequentially. This adsorbed metal-alkyl species may have an even greater surface mobility at a given substrate temperature than its elemental counterpart. At higher substrate temperatures, finally, a similar re-evaporation process occurs to that of the elemental species. As a consequence, the growth rate and also the residual (carbon) impurity concentration depend on the substrate temperature and on the group-V/group-III flux ratio in a very complicated manner [7,8,41].

A distinct example for the differences in growth mechanisms between ES MBE and GS MBE and for our poor understanding of the latter is the epitaxy of $Ga_xIn_{1-x}As$ lattice matched to InP substrate [41]. The difference is exemplified by the variation of In concentration with substrate temperature. In ES MBE the In concentration remains constant until loss of the least stable binary compound (i.e. InAs) occurs at higher temperature. A similar behavior, but over a wider temperature range, is obtained for MO VPE. However, the composition vs temperature profile changes noticeably with the use of metal alkyls in GS MBE. At low temperature, the pyrolysis of the Ga alkyl is less efficient as compared to the In alkyl, resulting in an increased In content in the $Ga_xIn_{1-x}As$ alloy. When the substrate temperature is increased, the pyrolysis rates stabilize and lattice matching of the $Ga_xIn_{1-x}As$ with x = 0.47 is achieved, but only over a very narrow temperature range. When the temperature is further increased, the In concentration in the alloy increases rapidly. This trend is opposite to that observed in ES MBE and is not yet understood. Obviously, this change in alloy composition is caused by an anomalous enhancement of the Ga alkyl desorption in the presence of an incident In alkyl flux. This problem of competing surface reactions highlights our current poor understanding of surface chemistry involved in MO MBE and in CBE.

When the decomposition of the metal alkyls on the growing surface is not completed, carbon may be incorporated into the growing layer [8,40]. This tendency for the grown material to be heavily p-type is more pronounced with TMG as starting material. In the case of TEG the so-called Pt-elimination process facilitates the breaking of the Ga-carbon bond so that the background doping level in GaAs can be reduced to the low 10^{14} cm^{-3} range [8]. In general, starting metalorganics having a weaker bonding of the metal alkyl and allowing for Pt-elimination are advantageous for MO MBE. However, whether the residual carbon incorporates with the group III radical or from chemisorbed organic radicals is not yet clarified. Also in the various subcategories of gas-source molecular beam epitaxy RHEED and RDS are now more frequently used to monitor *in-situ* the surface chemistry and growth mechanisms underlying the epitaxial growth [29,41].

The various subcategories of GS MBE close the gap between the techniques of conventional ES MBE and low-pressure MO VPE, as indicated in Fig. 2. These more recent developments have advantages for the growth of III-V semiconductors containing phosphorus. Although a number of impressive results have been achieved, a detailed investigation of the incorporation behavior of phosphorus and arsenic in ternary III-V-V compounds during HS MBE and CBE is still lacking. In addition, a thorough comparison of the properties of Al containing heterostructures and superlattices grown either by ES MBE and HS MBE or by MO MBE and CBE would reveal the actual state of the art of each of these techniques. Finally, the real challenge for both GS MBE and MO VPE is the replacement of the extremely toxic AsH_3 by suitable safer arsenic compounds of high purity.

3.4 Modulated Beam Flow Techniques

In the last six years, the original methods of vapor phase and beam epitaxy have encountered important new developments to expand their application particularly to low-temperature and hetero-epitaxial (lattice-mismatched) growth. While in the conventional techniques, the molecular or metalorganic beams (flows) impinge continuously onto the substrate surface for the growth of a homogeneous layer (e.g. GaAs), alternating or modulated beams (flows) synchronized with the layer-by-layer growth mode by properly actuating the shutters or the mass flow controllers are now used in migration enhanced epitaxy (MEE) [42], in atomic-layer MBE (AL MBE) [43], and in flow-rate modulation epitaxy (FME) [44]. To a certain extent these modulated beam techniques artificially induce an *atomic* layer-by-layer growth sequence on the (001) surface. Their advantages are the superior smooth morphology of surfaces and interfaces especially for lattice-mismatched (strained) systems, the suppression of defect formation, and the substantial lowering of the favorable growth temperature for high-quality material.

4 Additional Lateral Patterning

Additional lateral patterning of artificially layered III-V semiconductors leads to a new class of exotic microstructures, including quantum wires (QWR) and quantum dots (QD). The lateral confinement to quantum well structures can be applied with the techniques of electrostatic squeezing, lithographically defined mesa-etching, focussed ion beam writing, or growth-dynamics control through textured substrates. The condition for the occurrence of new phenomena in these low-dimensinal semiconductor structures is that the lateral size of the active region can be made smaller than the coherence and elastic scattering lengths. In this regime the structures act as electron waveguides, because the "lateral" wire dimensions are of the order of the de Broglie wavelength and only a few laterally defined modes are occuped. In addition to the wave nature of the electron being fundamental to the phenomena under study, the density of states changes drastically from a parabolic curve for three-dimensional (bulk) semiconductors, to a staircase shape for quasi two-dimensional (2D) structures, to a saw-tooth shape for quasi-one-dimensional (1D) structures, and finally to a delta-function like behaviour for quasi-zero-dimensional (0D) structures.

The minimum lateral dimensions obtained by patterning of artificially layered structures are mostly larger than the vertical dimensions. This leads to narrow spacings of the subband energies only, which are often masked by the level broadening due fluctuations of the wire width and defects introduced during the patterning process. To reduce the defect density, several methods for direct fabrication of quantum-wire and quantum-dot structures based on epitaxial growth have been exploited including growth of tilted superlattices on vicinal substrates [45], growth on V-grooved substrates [46] and strain induced confinement [47]. In these structures lateral dimensions comparable to the vertical ones can be achieved. They allow in principal for large subband spacings which are required for optical and electrical device applications [48].

The method of MO VPE has been successfully applied to the direct epitaxial growth of buried GaAs QWR structures on patterned non-planer substrates [46]. This approach relies on the fact that $Al_xGa_{1-x}As$ layers grown on nonplanar substrates exhibit lateral thickness modulations, which translate into lateral bandgap modulations. MO VPE of $Al_xGa_{1-x}As$ in etched channels oriented along [011] on a GaAs(100) substrate results in sharp V-shaped grooves evolving during growth. Subsequent growth of GaAs layers in these grooves yields crescent-shaped GaAs QWRs formed by migration and accumulation of Ga at the bottom of the groove. This technique provides two unique advantages for the reproducible fabrication of high quality QWRs: (i) all QWR interfaces are formed *in situ* and (ii) the (re-)sharpening of the V-grooves during the Al_x-$Ga_{1-x}As$ growth is self-limiting, i.e. the resulting shape is independent of the initial surface profile. Efficient luminescence from these QWRs comparable to that from conventional quantum well structures has been observed [46].

The most established approach based on MBE is the growth of tilted superlattices by the deposition of fractional monolayers of alternating composition on stepped surfaces created by a small misorientation from singular surfaces. However, its successful application has so far been very limited due to poor control of local misorientation, kink formation and stability of the growth rate [45]. The produced wires therefore suffer from nonuniformity in shape, dimension and direction, and up to now no clear manifestation of one-dimensional (1D) confinement effects has been observed. To overcome these difficulties we have developed two new approaches to directly synthesize III-V semiconductor QWR and QD structures by conventional elemental-source molecular beam epitaxy (ES MBE). First, we have fabricated isolated InAs QDs of subnanometer size embedded in a crystalline GaAs matrix by the controlled deposition of fractional monolayers of InAs at the step edges of terraced (001) GaAs surfaces [49]. The interfaces of the strained InAs QDs are in registry with the surrounding matrix. The absence of any surface and interface states makes this system promising to study the intrinsic optical response of ultrasmall zero-dimensional systems. Second we have fabricated GaAs QWR

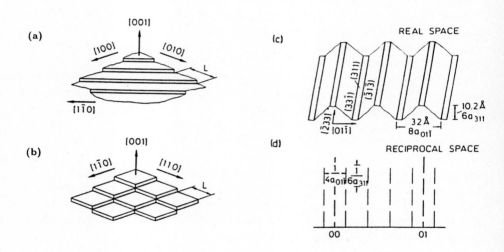

Fig. 9 (a) Terraces on vicinal (001) GaAs surface tilted towards [110]. (b) Terraces on vicinal (001) GaAS tilted [100]. In (a) and (b) the step height is one lattice plane (= 2.8 Å). (c) Scheme of stepped (311)A GaAs surface. (d) Reciprocal lattice (= RHEED pattern) of the stepped surface giving lateral periodicity and step height.

structures in an AlAs matrix by epitaxial growth of GaAs/AlAs multilayer structures on (311)A oriented substrates [50]. Periodic arrays of macro-steps composed of two sets of (331) and (313) facets are formed *in situ* having a periodicity of 3.2 nm and a step height of 1.02 nm. The as-grown QWR structures consist of ordered GaAs channels in the AlAs matrix extending along [233]. They show a distinct polarization dependence and an extremely high intensity of the luminescence even at room temperature. These periodic arrays of semiconductor quantum wires and quantum dots exhibit novel optical anisotropies and interband optical transitions, which allow to fully exploit these exciting properties in fundamental research and in photonic devices [48].

5 Characterization

The charcterization of semiconductors which are microscopically structured down to atomic dimensions requires analytical techniques having a high spatial resolution and a high sensitivity and accuracy. The necessity of a detailed characterization stems directly from the history of semiconductor superlattices [1]. First, a sound theory made it possible to predict the intriguing properties of these artificially layered semiconductors. Then, sophisticated measurement techniques were used to assess the degree to which the predictions have been fulfilled [2,3,4,11,12]. The interface perfection experimentally attainable by beam epitaxy is now so high that conventional methods of depth profiling analysis, which involve sputtering to section the material combined with some means of composition determination (such as Auger electron spectroscopy or secondary ion mass spectroscopy) are no longer adequate for the resolution required for atomically sharp interfaces. Hence, our ability to grow complex multilayer structures has led to new challenges for the characterization of materials, and experimental methods and their theoretical framework are developed to meet this challenge. Improved and new techniques to characterize materials have in turn a strong impact on the evolution of methods for spatially resolved materials synthesis.

In general several different charcterization methods are now used routinely to assess the specific structural and electronic properties of microscopically structure semiconductors: (i) transmission electron microscopy (TEM), (ii) high-resolution x-ray diffraction, (iii) emission and absorption spectroscopy, (iv) Hall effect and current-voltage or capacitance-voltage measurements, and (v) Raman scattering. In many cases, a combination of two or more of these methods is necessary for a clear assessment of the specific structural and electronic properties.

6 Conclusion

The improvement of epitaxial growth techniques has reached a status where monolayer dimensions in artificially layered semiconductor crystals are being routinely controlled to form a new class of materials with accurately tailored electrical and optical properties. The unique capabilities of the advanced epitaxial growth techniques in terms of spatially resolved materials synthesis has stimulated the inspiration of device engineers to design a whole new generation of electronic and photonic devices based on the concept of band-gap engineering. This concept, also called wavefunction or density-of-states engineering, respectively, relies on the arbitrary modulation of the band-edge potential in semiconductors through the abrupt change of composition (e.g. GaAs/AlAs, $Ga_xIn_{1-x}As/InP$, GaSb/InAs, Si/Ge, etc.) or of dopant species normal to the growth surface.

Epitaxial growth of custom-designed semiconductor microstructures consisting of alternating layers of chemically different materials having nearly equal lattice constants is routine now in many laboratories around the world. However, close lattice-constant matching is not a fundamental limitation for the growth of high-quality multilayer structures. For lattice mismatch as large as 7 % it is now possible to grow a finite thickness of epitaxial layer free of crystal defects. In this commensurate growth the lattice constant of the epilayer in the growth plane is strained to exactly match the lattice constant of the underlying substrate. As a result, the strained epilayer undergoes a tetragonal distortion, and the lattice constant in growth direction is no longer equal to the in-plane one. The ability to fabricate such strained-layer heterostructures is very attractive, not only because of the large variety of material combinations that can be produced, but also because of the use of the built-in strain to tailor the bandgap and the transport properties of such systems.

Applying additional lateral confinement to these artificially layered materials via submicron patterning has produced a new class of exotic semiconductor structures in which the quantum-mechanical properties of the electrons (holes) can be fully exploited. This microscopic structuring or engineering of semiconducting solids to within atomic dimensions is being achieved by the incorporation of interfaces (consisting of abrupt homo- or heterojunctions) into a crystal in well-defined geometrical and spatial arrangements. The electrical and optical properties of these low-dimensional semiconductor structures are then defined locally, and phenomena related to extremely small dimensions ("quantum size effects") become more important than the actual chemical properties of the materials involved.

Acknowledgement

This work was sponsored in part by the Bundesministerium für Forschung und Technolgie of the Federal Republic of Germany.

References

[1] L. Esaki and R. Tsu, IBM J. Res. Develop. **14**, 61 (1970)
[2] For a review on the first decade of MBE see: A.Y. Cho and J.R. Arthur, Prog. Solid-State Chem. **10**, 157 (1975)
[3] L.L. Chang and K. Ploog (Eds), Molecular Beam Epitaxy and Heterostructures (Martinus Nijhoff, Dordrecht, 1985), NATO Adv. Sci. Inst. Ser., Vol. **E87**, (1985)
[4] B.A. Joyce, Rep. Prog. Phys. **48**, 1637 (1985)
[5] G.B. Stringfellow, Organometallic Vapor Phase Epitaxy: Theory and Practice (Academic Press, Boston, 1989)
[6] G.B. Stringfellow, Mater. Res. Soc. Symp. Proc. **145**, 171 (1989)
[7] M.B. Panish and H. Temkin, Ann. Rev. Mater. Sci. **19**, 209 (1989)
[8] W.T. Tsang, in: VLSI Electronics: Microstructure Science, Ed. N.G. Einspruch (Academic Press, New York, 1989) Vol. **21**, p. 255
[9] S. Dushman; in: Scientific Foundations of Vacuum Technique, Ed. J.M. Laffarty, (John Wiley, New York, 1962) p. 80
[10] For a recent survey see: Proc. 3rd Int. Conf. Chemical Beam Epitaxy & Related Growth Techniques 1991 [CBE-3], Eds. G.J. Davies, J.S. Foord, and W.T. Tsang (Elsevier, Amsterdam, 1992), J. Cryst. Growth, Vol. **120** (1992)
[11] A.C. Gossard, Treat. Mater. Sci. Technol. **24**, 13 (1981)
[12] K. Ploog, Angew. Chem. Int. Ed. Eng. **27**, 593 (1988)
[13] For a recent survey see: Proc. 6th Int. Conf. Molecular Beam Epitaxy 1990 [MBE-VI], Eds. C.W. Tu and J.S. Harris (Elsevier, Amsterdam, 1991), J. Cryst. Growth, Vol. **111** (1991)
[14] C.T. Foxon and B.A. Joyce, Curr. Topics Mater. Sci. **7**, 1 (1981)
[15] M. Ilegems, J. Appl. Phys. **48**, 1278 (1977)
[16] D.L. Miller and P.M. Asbeck, J. Appl. Phys. **57**, 1816 (1985)
[17] For a recent survey on carbon doping of GaAs see: R.J. Malik, J. Nagle, M. Micovic, T. Harris, R.W. Ryan, and L.C. Hopkins, J. Vac. Sci. Technol. **B10**, 850 (1992) and references therein
[18] E. Nottenburg, H.J. Bühlmann, M. Frei, and M. Ilegems, Appl. Phys. Lett. **44**, 71 (1984)
[19] J. Maguire, R. Murray, R.C. Newman, R.B. Beal, and J.J. Harris, Appl. Phys. Lett. **50**, 516 (1987)
[20] L. Gonzales. J.B. Clegg. D. Hilton. J.P. Gowers. C.T. Foxon, and B.A. Joyce, Appl. Phys. **A41**, 237 (1986)
[21] For a review on delta-doping see: E.F. Schubert, J. Vac. Sci. Technol. **A8**, 2980 (1990)
[22] W.I. Wang, Surf. Sci. **174**, 31 (1986); H. Nobuhara. O. Wada, and T. Fujii; Electron. Lett. **23**, 35 (1987)
[23] R. Nötzel, L. Däweritz, and K. Ploog, Phys. Rev. **B46**, 4736 (1992)

[24] E.H.C. Parker (Ed.), The Technology and Physics of Molecular Beam Epitaxy (Plenum Press, New York, 1985); M.A. Herman and H. Sitter, Molecular Beam Epitaxy, Fundamentals and Current Status (Springer-Verlag, Berlin, Heidelberg, 1989), Springer Ser. Mater. Sci., Vol. 7 (1989)
[25] A.Y. Cho, J. Appl. Phys. 42, 2074 (1971)
[26] T. Sakamoto, H. Funabashi, K. Ohta, T. Nakagawa, N.J. Kawai, T. Kojima, and K. Bando, Superlatt. Microstruct. 1, 347 (1985)
[27] B.A. Joyce, P.J. Dobson, J.H. Neave, K. Woodbridge, J. Zhang, P.K. Larsen, and B. Bolger, Surf.Sci. 168, 423 (1986)
[28] B.A. Joyce, J. Zhang, J.H. Neave, and P.J. Dobson, Appl. Phys. A45, 255 (1988)
[29] D.E. Aspnes, IEEE J. Quantum Electron. QE-25, 1056 (1989)
[30] For a review on the first decade of MO VPE see: H.M. Manasevit, J. Cryst. Growth 55, 1 (1981)
[31] M.J. Ludowise, J. Appl. Phys. 58, R31 (1985)
[32] J.P. Duchemin, M. Bonnet, F. Koelsch, and D. Huyghe, J. Electrochem. Soc. 126, 1134 (1979)
[33] M. Razeghi, The MO CVD Challenge (Adam Hilger, Bristol, 1989) Vol. 1 (1989)
[34] A. Ishibashi, Y. Mori, M. Itabashi, and N. Watanabe, J. Appl. Phys. 58, 2691 (1985)
[35] N. Holonyak, Sov. Phys. Semicond. 19, 943 (1985)
[36] R. Luckerath, P. Balk, M. Fischer, D. Grundmann, A. Hertling, and W. Richter, Chemitronics 2, 199 (1987)
[37] M.E. Pemble, Chemitronics 2, 13 (1987)
[38] T. Makimoto, Y. Yamauchi, N. Kobayashi, and Y. Horikoshi, Jap. J. Appl. Phys. 29, L207 (1990)
[39] M.B. Panish, J. Electrochem. Soc. 127, 2729 (1980)
[40] E. Veuhoff, W. Pletschen, P. Balk, and H. Lüth, J. Cryst. Growth 55, 30 (1981)
[41] A. Robertson, T.H. Chiu, W.T. Tsang, and J.E. Cunningham, J. Appl. Phys. 64, 877 (1988); G.B. Stringfellow, Prog. Cryst. Growth Character. 19, 115 (1989); D.A. Andrews and G.H. Davies, J. Appl. Phys. 67, 3187 (1990); T. Martin, C.R. Whitehouse, and P.A. Lane, J. Cryst. Growth 120, 25 (1992); N.K. Singh, J.S. Foord, P.J. Skevington, and G.J. Davies, J. Cryst. Growth 120, 33 (1992)
[42] Y. Horikoshi and M. Kawashima, J. Cryst. Growth 95, 17 (1989)
[43] F. Briones, L. Gonzales, and A. Ruiz, Appl. Phys. A49, 729 (1989)
[44] Y. Horikoshi, K. Kawashima, and H. Yamaguchi, Jap. J. Appl. Phys. 27, 169 (1988)
[45] M. Tsuchiya, J.M. Gaines, R.H. Yan, R.J. Simes, P.O. Holtz, L.A. Coldren, and P.M. Petroff, Phys. Rev. Lett. 62, 466 (1989); T. Fukui and H. Saito, Jpn. J. Appl. Phys. 29, L731 (1990)

[46] E. Kapon, D. Hwang, and R. Bhat, Phys. Rev. Lett. 63, 430 (1989)
[47] D. Gershoni, J.S. Weiner, S.N. Chu, G.A. Baraff, J.M. Vandenberg, L.N. Pfeiffer, K. West, R.A. Logan, and T. Tanbun-Ek, Phys. Rev. Lett. 65, 1631 (1990); K. Kash, B.P. v.d. Gaag, J.M. Worlock, A.S. Gozda, D.D. Mahoney, J.P. Harbison, and L.T. Florez, in: Localization and Confinement of Electrons in Semiconductors, Eds. F. Kuchar, H. Heinrich, and G. Bauer (Springer-Verlag, Berlin, Heidelberg, 1990), Springer Ser. Solid-State Sci., Vol. 97 (1990)
[48] E. Corcoran, Scientific American 263, 74 (1990)
[49] O. Brandt, L. Tapfer, K. Ploog, R. Bierwolf, M. Hohenstein. F. Phillipp, H. Lage, and A. Heberle, Phys. Rev. 44, 8043 (1991)
[50] R. Nötzel, N.N. Ledentsov, L. Däweritz, M. Hohenstein, and K. Ploog, Phys. Rev. Lett 67, 3812 (1991); R. Nötzel, N.N. Ledentsov, L. Däweritz, and K. Ploog, Phys. Rev. B45, (1992)

Technology and Fabrication of Quantum Devices: Submicron Lithography and Etching Techniques

Sybrand Radelaar

DIMES, (Delft Institute for Micro-Electronics and Submicrontechnology)
Delft, The Netherlands

1 Introduction

In recent years many interesting physical effects in small structures have been discovered. The field goes by various names but we will use the term mesoscopic physics. The term was introduced to indicate that we are dealing with dimensions between the macroscopic and microscopic (i.e. atomic) regime. Typical dimensions in this regime range from 10 to 100 nm. Fluctuations due to the finite size play an important role in this range; it is therefore not surprising that the term was introduced first in stochastic processes (see e.g. van Kampen, 1981). However nowadays quantum effects, rather than fluctuations, form the main interest in structures with small dimensions.

This rapidly developing branch of solid state physics has been made possible by advanced fabrication techniques, which find their origin in semiconductor technology. The most advanced technology is used for the mass fabrication of memory devices like dynamic random access memories (DRAMs).

Up till about 10 years ago the practice in semiconductor laboratories and factories did not differ essentially. However, with the growing pressure to improve the yield of production, in particular for complex devices with more than a million transistors, industrial technology has grown tremendously in complexity and in cost.

The most advanced industrial semiconductor technology, which relies totally on *optical* lithography, has reached nowadays minimum feature sizes of the order of 0.5 μm in production and 0.35 μm in the development stage. This minimum feature size is in most cases insufficient for research in mesoscopic physics where, as mentioned above, one typically needs feature sizes of the order of 0.1 μm or less. For this purpose, special techniques like high resolution lithographic techniques were introduced and sophisticated pattern transfer techniques were developed. This particular combination of art and science is called *nanofabrication* or *nanotechnology*.

In this chapter we will discuss mainly two basic ingredients of nanofabrication techniques viz. high resolution lithography and etching. Other important aspects of the fabrication process like growth of semiconductor layers by MBE or CVD for control on an atomic scale in the vertical direction will be dealt with by other authors in this volume. We will, however, briefly touch upon some special techniques like selective growth which can, in special cases, be extremely useful for nanofabrication purposes.

This chapter is organized as follows: First, we discuss the most commonly used lithographic techniques useful for mass production and indicate why these techniques are only occasionally used for the fabrication of quantum devices. The most commonly used technique for nanolithography, viz. pattern definition in resist by direct write electron beam exposure, is described in some detail. Subsequently, we discuss the techniques that can be used to transfer this resist pattern into the substrate. The basic physical and chemical processes occurring during etching processes are reviewed. We conclude with a description of the fabrication of a few selected examples of devices.

2 Pattern Definition

2.1 The principles of device fabrication

The fabrication process of a structure consists, broadly speaking, of two essential types of processing steps viz. *pattern definition* (lithography) and *pattern transfer* (etching, selective deposition, etc.).

Lithography is the process by which a pattern is defined in a layer, the so-called *resist* layer, which is sensitive to irradiation by particles or photons. The resist layer becomes either soluble under influence of the irradiation, *positive* resist, or insoluble: *negative* resist. The basic sequence of a lithographic process step followed by pattern transfer, in this case etching, is illustrated in Fig. 1.

There exists quite a variety of techniques for the lithographic definition of patterns. The techniques can be divided into two main categories: *direct write* and *projection lithography*.

In the first case the pattern is written in the resist layer by scanning a beam of electrons or ions (in exceptional cases photons) over the resist layer. This type of technique is called "direct write". The process is *sequential*, i.e. the parts of the required pattern (pixels) are exposed or irradiated one after another.

In the second case the pattern is defined in the resist layer by projecting a mask onto the resist layer. Both photons (Deep UV or soft X-rays) or particles (electron or ion beam projection) can be used. The projection can be a simple 1:1 shadow image or a reduced image of the mask.

Techniques that make use of a mask, which must be fabricated by means of a "direct write" technique, offer the advantage that the mask can be used repeatedly. More importantly, the various parts of the image, the pixels, are exposed *simultaneously*, whereas for the direct write techniques the process is *serial*, pixel by pixel, and often extremely slow. For reasons of throughput optical projection lithography is the technique of choice for the electronic industry. For advanced research purposes time is less important because the number of devices is usually small and the pattern is often less complex. For quantum devices other criteria, like high resolution and flexibility, are much more important. For fabrication of these devices direct write electron or ion beam lithography is preferred, although the less critical parts of the patterns, like contact pads etc., are often defined with optical lithography tools. In this case the direct write tool must be aligned to the optically defined part of the pattern, This can be done with the aid of previously defined reference marks. This combination of more than one lithographic tool in one process is usually called mix and match technology. Ideally, optical and electron beam patterning is done in the same resist layer.

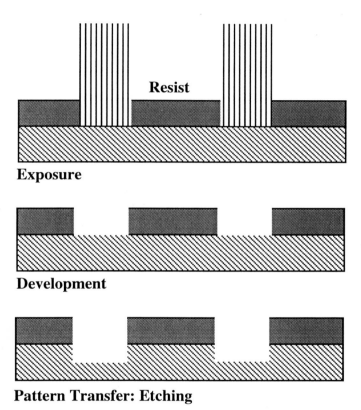

Fig.1   Basic steps in the fabrication of nanofabrication: pattern definition by means of exposure and development of resist and subsequent transfer of the pattern in the substrate by etching.

2.2 Optical Lithography

As mentioned above the semiconductor industry uses, almost without exception, optical lithography for the fabrication of integrated circuits. For high quality lenses the resolution of optical lithography is limited by diffraction. The ultimately achievable resolution is determined by the wavelength of the light and the numerical aperture of the projection system. The following criterion, due to Rayleigh, is often used to define the limiting resolution of an optical projection system:

$$\delta_{min} = k \frac{\lambda}{NA}$$

where λ is the wavelength of the radiation used, NA is the numerical aperture of the projection system and k is a constant of the order of unity (the actual magnitude varies somewhat with process conditions).

It is clear that the resolution improves with decreasing wavelength. However, the wavelength cannot be decreased indefinitely because there are several natural limits like the decreasing depth of focus:

$$\delta_f = k_2 \frac{\lambda}{(NA)^2}$$

or the fact that no suitable lens materials exist for very short wavelength radiation. Both equations clearly show that the use of optical lithography for nanofabrication is severely limited. At the moment a practical limit, using deep ultraviolet radiation, lies at ≈ 0.15 μm, which is too large for the greater part of the physical research of current interest.

2.3 X-ray Lithography

X-ray lithography has been investigated mainly for large scale production purposes. For 1:1 shadow projection, which is most common in X-ray lithography, the resolution is limited by penumbral shadow effects due to the finite size of the source and, for small source sizes (below 1 mm), by Fresnel diffraction. The limit of resolution in this case is approximately described by:

$$\delta_{min} \approx \sqrt{\lambda z}$$

where λ is the wavelength of the radiation (≈ 1 nm), and z is the distance between mask and substrate. The minimum distance is obviously set by the resist thickness, ≈ 100 nm, but this technology, called *contact printing*, often results in damage to the very fragile X-ray masks. In practice a small gap between mask and substrate is present: *proximity printing*.

One presently estimates that the practical limit of X-ray lithography using shadow projection lies somewhere around 50 nm. The real technological challenge is to produce defect-free masks with sufficiently low distortion. The problem is particularly severe because present day X-ray lithography is based on 1:1 shadow projection, which means that the line widths on mask and wafer are identical. If one wants to go to lower critical dimensions X-ray projection techniques based on reflection optics must be developed. Although up till now the use of X-ray lithography in nanofabrication has been limited, clever use of available resources has led to useful results in the fabrication of devices for mesoscopic physics. E.g. grating gate devices have been made using X-ray lithography (Warren et al).

2.4 Electron Beam lithography

2.4.1. Electron beam pattern generators

In industry electron beam lithography is mainly used for the production of masks for optical lithography and, to a limited extent, for direct write on wafers for application specific circuits (ASICs). Because of its great flexibility and high resolution, electron beam lithography is by far the most frequently used technique for nanofabrication. High resolution electron beam pattern generators are commercially available (e.g. Leica Cambridge, JEOL).

A schematic layout of an electron beam pattern generator (EBPG) is shown in Fig 2. An EBPG consists of an electron optical column which projects a reduced image of the electron source on the substrate; the resulting spot size is $\approx$ 8 nm. Smaller spot sizes can be achieved but it that case the resolution of the EBPG is likely to be limited by other effects e.g. the size of the resist molecules.

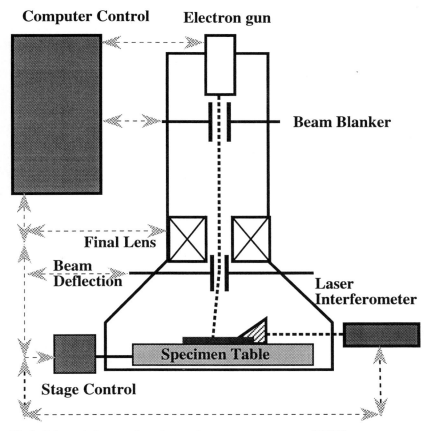

Fig 2. Schematic layout of an electron beam pattern generator (EBPG)

The electron beam can be switched on and off by means of a beam blanker. This is a simple device which can deflect the beam completely out of the optical axis. The main requirement of the blanker is speed. One of the main advantages of dedicated pattern generators is that they are equipped with laser interferometer controlled specimen stages with nanometer resolution, which allow automatic and accurate alignment of the pattern using previously defined reference marks. In this way multilevel structures can be produced. The electron gun is either a conventional tungsten or LaB_6 source or a field emission gun for the most advanced machines. The accelerating voltage can be varied, most EBPGs have power supplies that can go up to 100 kV.

The beam can be moved rapidly over the substrate by two sets of deflection coils, for movements in the x and y directions respectively. For advanced beam writers the frequency is typically 25 MHz. The maximum scan field is of the order of 1 mm^2. For larger patterns the scan fields are stitched by displacing the substrate by means of stepping motors. EBPGs offer the possibility to move the beam to a predescribed position, write a part of the pattern and move to the next position. This so-called *vector-scan* method prevents unnecessary movements of the beam position over areas where no exposure is needed and, for a given frequency, results in a shorter writing time compared with *raster-scan* systems, which move the beam across the substrate in the manner of the electron beam in a television tube. Usually raster scan systems allow a higher frequency which partly diminishes the advantage of vector-scan systems. For dense patterns the difference in performance is small anyhow.

A computer provides the dataflow for the control of the blanker, electron beam deflection and stage movements. The patterns are designed using CAD software and are subsequently translated into machine instructions by so-called post-processing software.

Many research groups have modified a Scanning Electron Microscope (SEM) for pattern writing by insertion of a beam blanker in the column. As the name implies, scanning electron microscopes move the beam in a raster-like fashion. This is of course a sound for idea for imaging systems but, as we mentioned above, this strategy is relatively time-consuming for patterns where only a small fraction of the area must be exposed. Modified commercial SEMs also are not equipped with laser interferometers and automatic alignment possibilities. Alignment of the next pattern in an SEM is done by first forming an image of the existing pattern and manually adjusting the position of the beam. For commercial SEMs the maximum accelerating voltage is usually limited to 25 or 30 kV. This is a particularly unfortunate choice from the point of view of electron beam lithography because the proximity effect is most severe in this range (see section 2.4.2 below).

Some workers have adapted a scanning transmission electron microscope (STEM) for nanolithography. An advantage of this type of microscope, which is normally used to make images of transparant substrates, have much higher accelerating voltages, up to 350 kV, are available. This can greatly reduce proximity effects (see sect.2.4.2). The disadvantage in comparison with an EBPG is the fact that only very small substrates can be used and precise stage control is not available.

The ultimate resolution of electron beam lithography with conventional resists is limited by the size of the molecules of the resist ($\approx$ 5nm) and the range of secondary electrons ($\approx$ 5 nm) produced by inelastic collisions of the impinging energetic electrons in the resist.

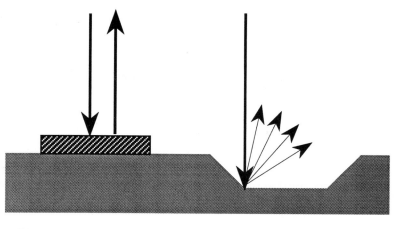

Heavy metal mark **Topographical mark**

Fig. 3. Different types of marks used in e-beam alignment. Heavy metal marks rely on the enhanced backscattering with respect to the substrate, etched marks change the backscattering by local modifications in topography.

2.4.2 Proximity effects

One of the most annoying problems in electron beam lithography is the so-called *proximity effect* which is caused by electrons which penetrate the resist layer and are backscattered from the substrate into the resist. These electrons form a haze around each feature of the pattern that has to be written. The dose received by a given part of the pattern depends on the dose and geometry of adjacent parts. For the case where a substantial fraction of the incoming electrons penetrate the resist layer and end up in the substrate the dose d(r) at a given position r on the pattern is given by:

$$d(r)=k\{\exp(-r^2/\beta_f^2)+\eta(\beta_f^2/\beta_b^2)\exp(-r^2/\beta_b^2)\}$$

where β_f is the forward scattering range in the resist, β_b is the backscattering range and k and η are constants. The severity of the proximity effect is strongly dependent on the atomic number of the substrate. Intricate patterns on heavy metals like gold are very difficult to deal with properly. The same holds to a lesser degree for III-V semiconductors, at least in comparison with silicon. The double Gaussian dose distribution given above breaks down if a substantial fraction of the electrons end up in the resist layer.

Various strategies have been developed to circumvent the problem related to the proximity effect:

• *Use of high energy or low energy electrons.*

Electrons with an energy which is much larger than necessary to penetrate the resist layer will be scattered over a much wider range. Since the backscattered electrons reenter the resist layer over a much wider area, the local dose received by a given part of the pattern is less sensitive to the precise geometry of the surrounding parts of the pattern. A disadvantage of this solution is that the resists are less sensitive to high energy electrons and thus a higher dose is needed to fully develop the resist. E-beam pattern generators with 50 kV accelerating voltages are available and the next generation will be equipped with 100 kV capability.

The proximity effect can also be reduced by using low accelerating voltages, typically 1 kV. Because of the limited distance which low energy electrons can travel in resist the imaging layers must be very thin. Multilayer resist systems are necessary to provide sufficient protection of the substrate (see sect.2.6.2).

• *Dose and shape corrections*

It is possible to adjust the dose received by a given feature by calculating the total dose, correct for the dose received via the backscattering, calculate the influence of this correction on the surroundings, etc. For complex patterns this procedure is very time-consuming and requires considerable computing power. Commercial software packages are available. Neural network type dose adjustment procedures have also been worked out. The vector-scan systems offer added flexibility for proximity corrections because the dose can be readily adjusted by the beam stepping rate and hence the exposure dose for each element of the pattern separately. Raster-scan corrections require a second compensating exposure step.

A very similar approach is to adjust the shape of adjacent features. The technique can be combined with dose corrections. In fact the shape correction technique could be considered as an extreme case of dose correction.

• *Thin film substrates*

A clever way to avoid the proximity effect is to use very thin substrates like membranes of silicon nitride which are very strong. The membranes must be so thin that most electrons which pass through the resist are also able to pass through the membrane. In this way backscattering is almost completely suppressed.

This technique for fabrication of thin membranes on silicon wafers is now well established. This remedy for the proximity effect is of course not without its problems and cannot be universally applied. The membranes cannot be made too large otherwise there is to much distortion due to stresses built in the membranes and of course the fact that there is no substrate left, precludes the fabrication of active devices like transistors. Researchers have used this technique succesfully for the fabrication of passive metallic devices like ring structures for the study of the Aharonov-Bohm effect.

2.4.3 New approaches

Two interesting new approaches to electron beam lithography have recently been introduced viz. e-beam lithography with STMs (scanning tunneling microscopes) and miniature e-beam machines. Both techniques have high resolution capabilities and could in principle become considerably cheaper than the currently available sophisticated e-beam pattern generators.

Scanning tunneling microscopes have been in use now for some time. In the imaging mode the gap between a very sharp tip and the substrate is so narrow that electrons can tunnel from the tip to the substrate or vice versa. Basically the same instrument can be used for direct write if, instead of operating the tip in the tunneling mode, the tip is used in the so-called *field-emission mode*. In this latter case the distance between tip and substrate is too large to allow significant direct tunneling between tip and substrate. However, under high vacuum conditions the field strength at the tip can be made so high that the barrier for emission into the vacuum is narrowed to such an extent that electrons can tunnel through this barrier into the vacuum. In this way a high brightness electron source in close vicinity to the substrate is created.

An advantage of an STM type electron source close to the resist is that the electrons have very low energies in comparison to the energies used in conventional electron beam lithography. This greatly reduces the proximity effects because backscattered electrons do not have sufficient energy to re-enter the resist, but it also means that very thin resist layers must be used. Definition of nm lines in positive (PMMA) and negative resists have been demonstrated (see e.g. Marrian et al. 1992). The substrate must be conducting. This also holds for the resist, but this appears not be a very heavy constraint for the very thin resist layers used in STM-probe lithography.

Another interesting approach is the use of STM-type field emission sources in miniature electron beam columns. It can be demonstrated that lens aberrations are greatly reduced if one reduces the size of the electron optical system. Chang and co-workers are building prototypes to demonstrate the feasibility of this concept.

It has recently been demonstrated that STM tips can be used to manipulate atoms adsorbed on silicon substrates. This opens up the way for nanofabrication with *atomic resolution*. The subject is now actively pursued by various groups.

2.5 Ion beam lithography

A practical alternative for electron beam lithography is ion beam lithography, for which commercial machines are also available. In principle ion beam lithography has one great advantage over electron beam lithography viz. the absence of a significant proximity effect. This is due to the fact that ions, because of their much greater mass, do not scatter as easily as electrons with the result that backscattering of the impinging particles is virtually absent. Another advantage often quoted for ion beam lithography, the high sensitivity of resists for ion irradiation, is a blessing in disguise because for very small feature sizes the number of ions capable of fully developing the resist is so small that shot noise effects become important. Disadvantages are smaller beam currents (partly compensated for by higher resist sensitivity), contamination by the ions (most ion beam machines use liquid metal ion sources) and, for the time being, the somewhat lower resolution of ion beam systems (spot size ≈ 30 nm).

The resolution is most probably limited by the same factors that determine the ultimate resolution of electron-beam lithography viz. the range of secondary electrons produced by the incoming ions and the size of the resist molecules.

2.6 Resist systems

2.6.1 Conventional resist systems

An essential ingredient of pattern definition is the resist layer. The most commonly used resists consist of polymerized organic materials. A typical dose required for full development of sensitive resists is of the order of 10 $\mu C.cm^{-2}$. For PMMA the required dose is roughly an order of magnitude larger. The resolution is in principle limited by the size of the polymer molecules ($\approx$ 5 nm) and the range of secondary electrons. There has been some work on inorganic resists, but they have not become very popular because of the extremely high dose required (≈ 1 $C.cm^{-2}$).

There are two types of resists: positive and negative. A positive resist becomes soluble in a suitable solvent after irradiation, a negative resist insoluble. The mechanisms differ for different resists, in positive resists chain scission of the polymers reduces the molecular weight which result in a higher solubility. In negative resists the electron irradiation produces a radical which induces cross-linking. In general negative resists have a somewhat lower resolution. Another problem associated with negative resists is the swelling in the solvents used to remove the non-irradiated resist material. For nanofabrication one tries of course to use high resolution resists. A very popular high resolution resist is PMMA (polymethylmetacrylate). However this resist has a rather poor etch resistance. This problem can be circumvented by making use of multilayer resist systems, see below.

2.6.2 Multilayer resist systems

The stringent conditions imposed by nanofabrication: high resolution and perfect anisotropy are very difficult to fulfill with one resist layer. As is clear from the description given above, the resist layer has to fulfill a double task:

(i) as the medium for the definition of the pattern which requires some chemical modification of the material due to irradiation with particles or photons, and

(ii) protection of the underlying material during the sometimes very harsh conditions during pattern transfer.

It is clear that these two requirements are somehow contradictory, e.g. bombardment by energetic particles also occurs during etching and ion implantation, and therefore a compromise between sensitivity and the protection function must be found. A way to circumvent this compromise is to separate the functions of pattern definition and the protective (resist) function. For this purpose so-called *multilayer* resist systems have been developed.

In the preceding paragraph it was already indicated that the needs for high resolution in the imaging process and sufficient mask protection in the pattern transfer process are contradictory. The thinner the resist layer the higher the resolution but the less latitude remains to obtain the required etch depth. In addition there is also the strong requirement of dimension control. In particular when high aspect ratios (depth to width ratios) are necessary it is of utmost importance to ensure that the initial resist walls are as perpendicular as possible. A tapered mask profile will result in etched features which are also tapered if the selectivity and anisotropy of the used etch processes is not optimal.

The practical solution to circumvent these problems is Multilayer Mask Technology. In nanofabrication MMT technology is mandatory. The MMT process includes two essential features:

- Separation of the imaging and masking layer;
the imaging layer can be as thin as possible to preserve the highest possible resolution. This makes it also possible to use low energy electrons.

- Fully anisotropic pattern transfer from imaging to masking layer;
a sequence of plasma processing steps provides a fully anisotropic mask profile, the process is at low pressures to ensure the required anisotropy.

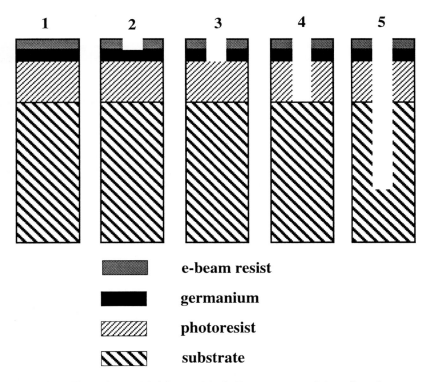

Figure 4. Three layer Multilayer Mask Process comprising 1) e-beam exposure, 2) development and descum in oxygen plasma, 3) dry etching of Ge in a fluor plasma, 4) anisotropic dry etching of bottom photoresist layer in low pressure oxygen plasma and 5) dry etching into the substrate layer.

The principle for a 3 layer positive tone mask is illustrated in Fig. 4. The thin imaging layer on the top and the bottom mask layer are separated by a thin (10-15 nm) germanium layer which is necessary to guarantee the pattern transfer from top to bottom layer. The crucial step is the anisotropic etch into the bottom layer by a low pressure oxygen (O_2) reactive ion etching (RIE) plasma (see below).

In the negative variant an even simpler scheme exists in which the top and intermediate layer are combined. This can be achieved by using negative e-beam resists which contain a certain amount of silicon. After e-beam exposure and development a low pressure oxygen plasma is applied. The oxygen reacts with the silicon in the remaining resist and forms a thin SiO_2 layer in the exposed regions. These thin oxide layers serve as a mask for subsequent anisotropic etching into the bottom layer. The positive variant of this double layer resist process is under development.

The ultimate simplification in MMT is *surface imaging*, a process in which only a single layer of resist is used. After exposure the resist reacts selectively, i.e. only in exposed regions, with a silicon containing molecule, which is chemically bonded in the surface layers of the resist. The latter process is called *silylation*. The process described above gives a negative tone mask, but (more complicated) positive tone alternatives have been reported.

The silylation can be done in a liquid or a gas. Treatment of the sylilated resist in an oxygen plasma locally forms a SiO_2-film which serves as a (negative) mask for subsequent anisotropic pattern transfer. The surface imaging process is primarily attractive in advanced optical lithography schemes in production processed where the loss of focussing depth in area with large topography can be circumvented by limiting the exposure to the top surface. The ultimate resolution in the sylilation process will be determined by diffusion of the Si-containing species. In state of the art megabit processes with finest details down to 0.35 µm the resolution will be sufficient. It may be doubtful whether in nanofabrication this process will be appropriate.

Multilayer masking can be used to create an undercut profile for lift-off applications or for shadow mask evaporation in ultra small Josephson junction array fabrication processes, see below. A variety of multilayer mask schemes have been reported in a recent review, where particular attention which given to e-beam lithography and ultimate resolution (Romijn et al.).

3 Pattern Transfer

Once the latent image has been transferred into the resist layer by development the pattern must be transferred into the substrate. One can distinguish again two basic types of pattern transfer viz. *subtractive* techniques in which one removes part of the underlying material and *additive* techniques in which material is deposited on the parts of the substrate which are not covered by the resist layer. An overview of the various pattern transfer techniques is given in Table 1. It is the author's opinion that workers in the field do not yet sufficiently exploit the potential of selective deposition techniques.

Table 3.1.

ADDITIVE TECHNIQUES	SUBTRACTIVE TECHNIQUES
lift-off	sputtering
selective deposition	reactive ion etching (RIE)
local ion implantation	plasma etching
	wet etching

3.1 Additive techniques

3.1.1 Lift-off technique

One of the additive techniques that is quite popular with researchers because of its simplicity is the so-called lift-off technique. It allows the fabrication of small structures which are composed of materials which cannot, or not easily, be etched by RIE or similar techniques. Another advantage is that high resolution resists like PMMA can be used. These types of resists usually have a very poor etch resistance.

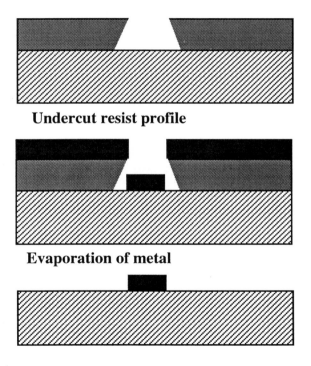

Undercut resist profile

Evaporation of metal

Lift-off of resist and metal

Fig.5 Principle of lift-off technique: metal is deposited on an undercut resist profile. The metal on top of the resist is floated off when the resist is dissolved.

The principle is explained in Fig5. The essence of the method is to create an overhanging resist profile i.e. the opening in the resist is narrower at the top than at the bottom. One way to achieve this is by means of a double layer resist system e.g. a two-layer PMMA resist which consists of a top layer with a high molecular weight and a bottom layer with a somewhat lower molecular weight. For an equal irradiation dose the rate of solution of the high molecular weight resist in the developer is lower than the bottom layer. The metal is subsequently deposited, usually by evaporation. Other techniques like sputter deposition can not be used because the trajectories of the incoming atoms must be perpendicular to the substrate. Because of the overhanging resist structure the material deposited on the substrate and on top of the resist are not in contact. If the remaining resist is removed by a suitable solvent the metal deposited on the resist disappears and only the metal on the substrate remains.

3.1.2 Selective deposition

As mentioned above lift-off is a suitable technique for the fabrication of small metallic structures which relies on *physical* vapour deposition methods like evaporation, which are not selective i.e. the material will be deposited on the patterned substrate irrespective of the chemical nature of the exposed material. *Selective* deposition of both metallic and semiconductor materials is possible by *chemical* means. The selectivity is achieved by the catalytic action of a part of the substrate. E.g. tungsten can be deposited on silicon from a gas like WF_6 and, under the same conditions, will not deposit on silicon oxide. A major advantage of the selective techniques is that the patterns are self-aligned: no additional lithographic patterning of the deposited layer is required. Useful selective techniques are:

• selective deposition from the gas phase (possible with Chemical Vapour Deposition (CVD) or Gas Source MBE);
• selective deposition from the liquid phase (electroplating or electroless deposition);
• selective reaction with the substrate material.

An example of the latter case is the reaction of titanium with silicon to $TiSi_2$. If the silicon is partly covered with an oxide titanium will only react with the exposed silicon. After reaction the remaining titanium can be removed by a suitable etchant which does not attack the silicide.

3.2 Subtractive techniques

3.2.1 Sputtering

The simplest technique that allows the removal of material under vacuum conditions is sputtering where ions with energies of a few hundred eV and up impinge on the substrate. The atoms of the substrate are knocked away by these energetic particles. One must resort to this technique if no suitable gas mixtures for the dry etching of a given material can be found. This is for example the case for gold. Unfortunately with physical sputtering techniques there is very little selectivity and the organic resist layers often erode much faster than the substrate material. For this reason very thick organic resist layers or a multilayer resist with an intermediate layer which is very resistant to physical sputtering must be used.

The conclusion is that sputtering is to be avoided at all costs. The reason to include sputtering in the list of subtractive techniques is that physical sputtering to some extent always take place in etch plasmas. The message is that also in etch plasmas great care has to be taken to avoid excessive sputtering.

3.2.2 Etching

The need for dry etching techniques

Etching is one of the most important and most difficult techniques used in the fabrication of nanostructures. As will demonstrated below the degree of anisotropy of the etching process is of paramount importance for nanofabrication. In the early days of semiconductor fabrication etching of material was mostly done with wet chemicals, but this method has now been replaced by so-called dry-etching techniques. The reason for this is simple: with only a few exceptions wet etchants are isotropic i.e. the etchants remove material at the same rate in all directions. This results in an increase of line-width, which is of the same order as the thickness of the material that has been removed. It is clear that for nanofabrication, where the conditions are even more severe than in memory production, wet etching cannot be accepted.

Isotropic etch   **Anisotropic etch**

Fig. 6 Pattern transfer by isotropic and anisotropic etching. Note that the increase in linewidth during isotropic etching is twice the depth to which the substrate is etched.

To satisfy the need for accurate line-width control during pattern transfer anisotropic dry etching techniques were developed. Dry etching is defined as the removal of material with the aid of low temperature glow discharges. The difference between sputtering and dry etching lies in the interplay between chemical and physical processes. By a proper choice of etch gases, radicals (reactive molecular fragments) can be created which selectively attack the material to be removed, whereas the ion bombardment induced by the plasma induces the required anisotropy. The degree of anisotropy can be tuned by proper choice of etching conditions like pressure, power etc. Note that plasma etching processes are not necessarily anisotropic. E.g. the stripping (removal) of resist after a pattern transfer step is usually done in so-called after-glow oxygen plasmas, where the removal rate is nearly isotropic. Here the resist is removed by the oxygen atoms produced in the plasma, the wafer is positioned in such a way that it is not subjected to ion bombardment.

Etch plasmas

A plasma is defined as a quasi-neutral gas of charged and neutral particles which exhibits collective behaviour (Chen 1977). As the term quasi-neutral indicates the plasma consists of equal concentrations of positive (ions) and negative particles (electrons). Electric fields are effectively shielded by the plasma. The screening length, the so-called Debye length, is given by

$$\lambda_D = \sqrt{\frac{\varepsilon\, kT}{e^2 n_0}}$$

where ε is the dielectric constant of the plasma, k Boltzmann's constant, T the absolute temperature, n_0 the electron density and e the electron charge.

For a typical glow discharge the Debye length is of the order of 1 millimeter, much smaller than the typical distance between the electrodes in a plasma etcher (a few centimeters).

The degree of ionization, defined as the ratio of the ion and neutral gas particle density, is rather small for low pressure (1-200 Pa) rf-plasmas : of the order of 10^{-4} to 10^{-6}. This concentration of ions might seem very low, but one must realize that this degree of ionization is accompanied by a much higher concentration of dissociated molecules (radicals), typically two to three orders of magnitude larger.

The energy distributions of the particles in a plasma deviate from the Maxwell-Boltzmann distribution, this is particularly true for the electrons. Nevertheless, the energy distribution of the electrons is often qualitatively described using only one parameter, the electron temperature. This succinct description poses problems because processes like ionization and dissociation are very sensitive to the high energy tails of the electron energy distribution. The electrons and ions are not in thermodynamic equilibrium. Because the energy transfer between electrons and neutrals or ions is very inefficient, due to the large difference in mass, the average kinetic energy of electrons in a typical glow discharge is much higher than of the neutrals and ions.

A schematic view of a single wafer etcher is shown in Fig.7. The etcher consists of a vacuum chamber with two electrodes, one of which is usually grounded. The other is connected to a high frequency power supply. The etch gases and reaction products are very corrosive, therefore chemically resistant pumps have to be used. A variable conductance, usually a diaphragm valve, is mounted in the exhaust line to enable independent variation of pressure and flow rate.

The use of rf-discharges allows the etching of insulating materials like oxides and nitrides. The frequency at which these reactors operate is usually is 13.56 MHz, a frequency reserved for industrial use. At these high frequencies the ions in the plasma, contrary to the much lighter electrons, can no longer follow the rapid changes in the field direction. Shortly after the start of the discharge, electrons charge all surfaces surrounding the plasma (electrodes and surrounding walls) become negatively charged with respect to the plasma. After a few cycles the plasma potential saturates. The potential difference between electrodes and plasma ions are extracted from the plasma and bombard the electrodes. The electrons charge the electrodes during a short part of the rf-cycle.

Etch mechanisms

The etch mechanisms are often very complicated. The essence is however that *volatile* reaction products are formed which desorb and are subsequently removed via the pumps. The reactive species present in the plasma adsorb on the surface, a reaction takes place and the reaction product must be removed. The function of the ion bombardment is to enhance desorption of reaction products and (or) to enhance the reaction of radicals with the substrate material. The ions impinge more or less vertically on the substrate, depending on the number of collisions in the dark space and hence on the pressure, and in this way create the required anisotropy.

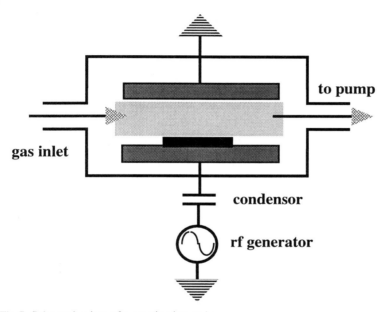

Fig 7. Schematic view of a reactive ion etcher.

The wafer is usually placed on the driven electrode; due to the fact that the opposite electrode is grounded and therefore connected to the wall of the reaction chamber the capacitance of the driven electrode is much smaller and the voltage drop across the dark space in front of the driven electrode is largest. The ions which impinge on the substrate material are responsible for the directionality of the etch process. In the absence of ion bombardment etching is virtually isotropic.
From this simple picture it is clear that two classes of particles produced by the plasma are very important: reactive neutrals (radicals) and positive ions. The *reactive neutrals* hit the substrate more or less random. Two types of chemical reactions involving these particles can take place:

-formation of volatile compounds.
   Evaporation of these compounds is responsible for the material transport from substrate to plasma and eventually to the pumps.

- production of surface coatings;
    Often a polymeric residue is formed on the side-walls of the groove which protects the walls against attack by the radicals.

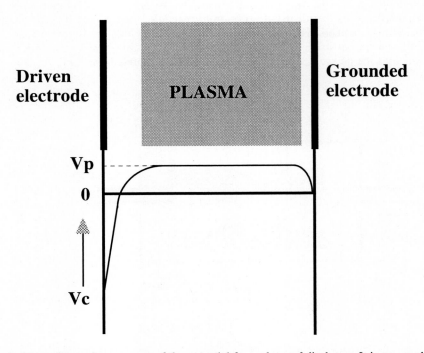

Fig 8. Approximate time-average of the potential for a planar rf discharge. It is assumed that the wall area is much larger than the area of the powered electrode.

The *positive ions* hit the substrate more or less perpendicularly if their mean free path is larger than the sheath (dark space) thickness. They are accelerated by the electric field in the dark space (plasma sheath) between the driven electrode and the plasma. Because the ions have relatively high energies, of the order of a few hundred eV, they can:

- locally enhance the reaction between substrate and radicals;

- remove material by physical sputtering;

- suppress surface polymerization e.g. by cutting the polymer chains.

Recent work has shown that apart from the ion flux there often is a significant flux of *energetic neutrals* which are produced by charge exchange between ions and neutral atoms in the dark space above the electrode (see e.g. Manenschijn et al.). This effect occurs already at pressures of about 50 Pa. The density of neutrals is much higher than the ion density and therefore ion-neutral collisions are still fairly frequent even if the ion concentration is as low as 10^{-5} (the mean free path is of the order of a few millimeters). The energy of these neutral particles is lower than the ions. Even in the absence of chemical reactions between neutrals and the resist layer, ions and energetic neutrals erode the resist layers by physical sputtering. This might cause loss of line width control and loss of anisotropy.

Requirements for the etching of nanostructures

For the fabrication of nanostructures two parameters are of utmost importance: *selectivity* and *anisotropy*. The reasons for the latter have already been explained in the introduction to this section. Selectivity is important because in general only the layer not protected by the resist must be removed. Therefore the etch process must be selective both to the resist and the underlaying layer. A high selectivity with respect to the resist is very important in nanofabrication because the etched profile will change if the resist is attacked. Unfortunately, during the development of an etch process compromises must be made since the two parameters cannot controlled independently.

The selectivity of an etch process is usually defined as the ratio of vertical etching of areas which should not be removed, like resists or an underlying layer, to the etching rate of the material to be etched. For practical work a selectivity of at least 1:10 is required. However if the underlying layers are very thin like e.g. 20 nm gate oxides selectivity must be considerable better.

For a given combination of materials the selectivity depends on a number of factors:

• reaction rate of the radicals with the various exposed materials;

• sputtering rates of the exposed materials;

• formation of passivating layers by preferential polymerization.

Selectivity is achieved if the composition of the plasma is adjusted in such a way that only the exposed part of the substrate layer is attacked by the neutral radicals. This is not always easy to achieve. An alternative strategy is to protect the other materials by means of a polymer layer. These passivating polymer films, which are sometimes hard to remove, are essential for the anisotropy of many industrial etching processes. However, if possible, the use of this principle should be avoided in nanofabrication because the side-wall polymers make the precisely controlled etching of small structures very difficult. Polymer films (with up to 0.1 µm thickness) can easily take up a substantial fraction of the width of the etched trench. With the continuing down-scaling of industrial devices like memories this has become also a problem in industrial processing.

Other etch parameters, like etch rate and uniformity over large areas, which are important in an industrial environment are less important for nanofabrication purposes; production time is seldom an issue and the samples used in nanofabrication are either small or only a few working devices on a wafer are required for the experiments. A problem that arises when one goes to very small dimensions is the so-called *microloading effect*.[1] In a very narrow structure diffusion of reactive species and reaction products is limited. The etch rate at the bottom of a very narrow trench is therefore lower than at the top.

Etch chemistry and selectivity

For the formation of passivating layers chlorofluorocarbons (CFCs) have been in use[2]. An example is the use of CF_4 for the etching of silicon oxide and silicon. In a CF_4 gas discharge one typically finds CF_3^+, F^+, F^-, CF_2^+ ions, F, CF_2 and CF_3 radicals and of course electrons. The CF_2 radicals formed in the plasma can combine to form a $[CF_2]_n$ polymer. However the plasma contains free fluorine radicals. These radicals limit the formation of the polymer by recombining with CF_2 to CF_4. If hydrogen is added to the gas fluorine atoms are scavenged by the formation of HF molecules and polymer formation can proceed. A few percent of hydrogen is sufficient to eliminate side-wall etching. One must keep in mind that the ion bombardment also impairs the formation of the polymer. Ion bombardment on the side-walls is negligible, however. In industry CHF_3 with small additions of more fluor rich compounds like or C_2F_6 diluted in a noble gas like Ar or He is most commonly used for the selective etching of silicon oxide. Note that the passivating layer principle works both to improve selectivity and anisotropy.

Aluminium can be etched in a chlorine or bromine atmosphere notwithstanding the fact that the volatility of solid aluminiumchloride is rather low: a vapour pressure of ≈100Pa near room temperature. The reaction with the native oxide does not start spontaneously but must be activated by ion bombardment. For anisotropy sidewall passivation is required.

One way to avoid the use of passivating films is to cool the samples during etching. The idea behind this principle is that at low temperature the evaporation rate of the volatile compound is reduced to such an extent that only on the regions where ion impact takes place the volatility will be sufficient. Good temperature control can be achieved by cooling the backside of the substrates with helium. For small samples which are often used in laboratories this approach is not very practicable.

Etch damage

An issue which can be very cumbersome is the damage caused by ion impact on the underlying substrate material. Energetic particles which are always present in reactive ion etching can cause degradation of the physical properties of the device. For example the electron mobility of 2–dimensional electron gases can be greatly affected by etch processing. Other effects are surface depletion in GaAs, creation of trapping centres near the silicon/ siliconoxide interface and the charging of oxides in

---

[1] In etching of large areas the loading effect denotes the decrease in etch rate with increasing exposed wafer area which is caused by the depletion of reactive species.

[2] In the year 2000 the use of CFCs will be banned because of their influence on ozone depletion in the atmosphere. Therefore the industry is actively trying to eliminate these gases from the workplace.

MOS structures. The effects are often more severe in III-V materials than in silicon or SiGe.

An additional cause for damage is the ultraviolet radiation produced by the recombination in the plasma and on the walls. Energetic photons can create charges in the oxide layers, which in turn influence the threshold voltage of the device. Often etching must be followed by a low temperature annealing step to remove the radiation damage. In order to diminish damage, equipment manufacturers have developed new approaches to etching technology. The obvious solution is to go to lower ion energies. Since the ion energy and the power supplied to the plasma are correlated, lower power almost always results in lower etch rates and loss of anisotropy.

In recent years much attention has been given to etching in electron cyclotron resonance (ECR) plasmas. The plasma is created in a microwave cavity. The frequency is 2.45 GHz a frequency available for industrial use. The cavity is surrounded by magnets which create an axial magnetic field which diverges slightly towards the substrate. The electrons in the plasma, following the flux lines, are driven out of the plasma cavity. The ions will follow this cloud of electrons to maintain charge neutrality.

The energy of the ions is substantially lower than the energy of ions in RIE etchers, sometimes so low that an additional substrate bias has to be supplied. This greatly reduces the radiation damage by plasma particles. On the other hand the UV-radiation emitted by the plasma is much more intense, in particular at smaller wavelengths, i.e. in the deep UV range. This means that oxide degradation by charging can become quite severe.

It is not possible in the context of this chapter to treat the background of the etching process in greater depth. For a more detailed introduction to the subject the author refers the reader to the recent book by van Roosmalen *et al.*

4 Direct Patterning (resistless) Techniques

As has been mentioned above Scanning Tunneling Microscopes used in the field-emission mode offer the possibility to pattern resists. It is also possible to use the STM as an electron source to irradiate a metastable molecular adsorbate. The adsorbate e.g. a metal-organic or inorganic compound like WF6 (tungsten hexafluoride) decomposes under electron irradiation into a metallic residue and radicals which evaporate. Thin metallic wires have been produced this way. The technique is still in its infancy and much more work has to be done before it can be used as a reliable nanofabrication tool.
The principle is of course not limited to STM's. Similar experiments have been done with conventional electron or ion beam writers.

5 Examples

Simple passive structures, for the observation of electron interference effects like the Aharonov-Bohm effect, are most easily produced by means of electron beam lithography in high resolution positive multilayer resists followed by lift-off (e.g.Verbruggen et al). Fig.9 shows an example of a ring structure for the study of the Aharonov-Bohm effect. The ring diameter is ≈ 0.8 µm and the width of the ring is ≈ 41 nm. For a clear observation of this effect the ring must be as thin as possible. The extraordinary flexibility of electron beam writers is demonstrated in Fig.10 which depicts a Penrose tiling made by electron-beam lithography and lift-off (courtesy D.van Leeuwen, DIMES).

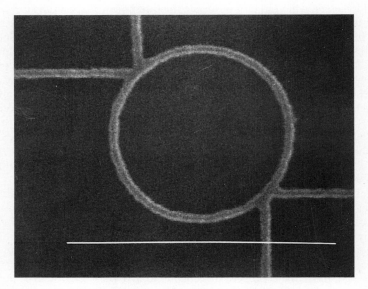

Fig. 9 Ring structure produced by electron beam lithography and lift-off.

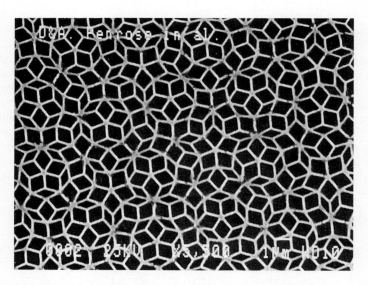

Fig.10 Penrose tiling produced by electron beam lithography and lift-off.

In the author's group thin membranes are often used for the fabrication of very small structures. As explained above the main reason is that the proximity effect is strongly reduced. Thin membranes are also used for the fabrication of so-called point contacts, very small contacts between two metallic or semiconductor layers, with diameters of the order of 10 nm. These contacts are used for the study of electron-phonon interaction etc.

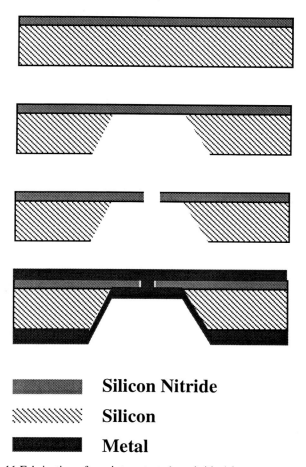

Fig.11 Fabrication of a point contact (nanobridge) between two metallic regions.

The first step is the fabrication of the membrane, which usually consists of silicon nitride. The nitride is deposited on a silicon wafer. Part of the material of the wafer is removed from the back side by means of wet anisotropic etching. The etch stops at the nitride layer. The next step is the etching of a small hole in the silicon nitride membrane. Metal is deposited from both sides of the hole onto the wafer and a thin bridge between the two metallic layer is formed (Holweg et al.). Similar techniques can be used to form nanobridges between semiconductors or metal/ semiconductor combinations.

6 Conclusion

Nanofabrication is a mixture of art and science. In general it requires a well equipped laboratory and patient people skilled in several techniques to be able to produce useful devices for fundamental research. However, often surprisingly good results have been obtained by inventive use of standard equipment and the application of clever, new tricks. Much useful material on device fabrication can be found in the proceedings of the Microcircuit Engineering Conference Series (published in the journal Microelectronic Engineering) and of the Electron, Ion and Photon Beam Conferences (published in the J. Vac.Soc.) Many useful references can be found in the paper on nanostructure technology by Chang et al. (Chang *et al.* , 1988)

7 Acknowledgements

The author wishes to thank his co-workers Dr.Ir.J.Romijn, Dr.A.H.Verbruggen and Dr.E.van der Drift, of the Section Submicron Technology of DIMES, for the critical reading of the manuscript and for the numerous stimulating discussions that we had in the past on various aspects of nanofabrication.

8 References

G.R.Brewer, Editor
   Electron Beam Technology in Microfabrication
   Academic Press, New York (1980)

Chang et al
   IBM J.Res.Develop. **32**, 462 (1988)

F.C.Chen
   Introduction to Plasma Physics
   Plenum , New York (1977)

P.A.M.Holweg, J.Caro, A.H.Verbruggen and S.Radelaar
   Fabrication of metallic nanoconstrictions
   Microelectronic Engineering **11** (1990) 27-30

C.A.van der Jeugd, G.J.Leusink, G.CAM. Janssen and S.Radelaar
   Appl. Phys. Letters **57** (1990) 354-356

D.P.Kern
   in Quantum Effects Physics, Electronics and Applications,
   K.Ismail,T.Ikoma and H.I.Smith editors
   Institute of Physics Conference Series Number 127, p. 73-78

A. Manenschijn , W.J.Goedheer
  Angular angular ion and neutral distribution in a collisional rf sheath
  J.Appl.Phys. **69** (1990) 2923-2930

C.R.K.Marrian and E.A.Dobisz
  High-resolution lithography with a vacuum STM
  Ultramicroscopy **42-44**, p.1309-1316 (1992)

E.A.Dobisz and C.R.K.Marrian
  Scanning tunneling microscope lithography: A solution to electron scattering
  J.Vac.Sci. Technol. **B 9** (1991) 3024-3027

P.M. Petroff et al.
  in Quantum Effects Physics, Electronics and Applications
  K.Ismail,T.Ikoma and H.I.Smith editors
  Institute of Physics Conference Series Number 127 (1992), p 85-94

J. Romijn and E. van der Drift
  Nanometer -scale lithography for large lateral structures
  Physica B **152** (1988) 14-21

A.J.van Roosmalen, J.A.G.Baggerman and S.J.H.Brader
  Dry Etching for VLSI
  Plenum Press, New York and London (1991)

A.H.Verbruggen, P.A.M.Holweg, H.Vloeberghs, C.van Haesendonck, J.Romijn, S.Radelaar and Y. Bruynseraede.
  Microelectronic Engineering **13**, 407 (1991)

A.C.Warren, D.A.Antoniadis, H.I.Smith and J.Melngailis
  IEEE Electron Device Letters EDL **6** (1985) 294

Band Gap Engineering in Low Dimensional Semiconductors

Roberto Cingolani

Dipartimento di Scienza dei Materiali, Universita' di Lecce, Italy

The advent of atomic-layer-controlled growth technologies, primarily Molecular Beam Epitaxy (MBE) and Metallorganic Chemical Vapor Deposition (MOCVD), has allowed the fabrication of macroscopic quantum-mechanical systems, in which the electronic properties are determined by quantum size effects. The capability of changing the gap of a semiconductor heterostructure by a suitable choice of the structural and compositional parameters, the so called *band-gap engineering*, has opened up new applications in the field of optoelectronics. In particular, low-dimensional semiconductors, namely quantum wells, quantum wires and quantum dots, have become privileged systems for the study of quantum confined electronic wavefunctions and for photonic appications. In this lecture we shortly survey the basic concepts of band gap engineering and its major applications.

1. **Basic theoretical concepts of electronic states and optical transitions in low-dimensional semiconductor heterostructures.**

In low-dimensional semiconductors the confinement of the electron and hole wavefunctions along one or more directions causes strong changes in the electronic energies and in the density of states dispersion. Electronic states along the free directions are described by plane-waves, whereas they are characterized by wavefunctions with the same periodicity of the heterostructure along the confinement directions. An effective quantum confinement can be achieved by growing a thin semiconductor layer in between two wider-gap semiconductor barriers. If the layer has a thickness comparable to the De Broglie wavelength of its electrons (about 10 nm) a two-dimensional quantum well is formed. If an additional barrier is put to laterally confine the carriers along one of the two remaining directions a one-dimensional quantum wire is obtained. The zero-dimensional limit (quantum dots or quantum boxes) is reached in small semiconductors volumes which are confined along all the three directions. The physical shape and the density of state dispersion of low dimensional semiconductors are depicted in Fig.1.

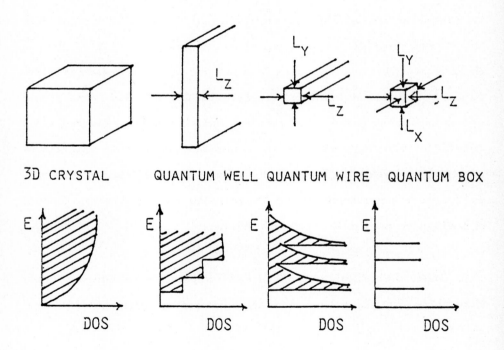

Fig.1 - Physical shape and density of states of low dimensional semiconductors.

There are different methods to calculate the electronic states confined in low-dimensional systems, which are based on very different conceptual approaches. In the following we will consider the quantum well case, in which carriers are confined along the growth direction. The extension to lower dimensionality systems can be straightforwardly performed taking into account the quantization energies along any additional confinement direction. A first method is the so called envelope function approach [1] , in which the periodic sequence of quantum wells (superlattice) is treated like a homogeneous crystal with a superimposed periodicity along the confinement direction. The envelope function varies slowly on the length scale of the lattice constant (a_o) and the crystal potential is taken into account by means of the effective mass approximation . The accuracy of the envelope function model is quite good in superlattices where the total period $d = L_z + L_b >> a_o$ (where L_z and L_b are the well and the barrier width, respectively). In this case the interface plane is really a minor perturbation on the slowly varing envelope function. The disadvantage of the envelope function approach in its simplest form lies in the applicability to energy states very close to the band gap edge of the constituent bulk materials. In addition , the accuracy of the model drops in ultrathin layer heterostructures , where the superimposed periodicity approaches the bulk crystal period ($d \simeq a_o$). In these cases microscopic models are needed , which treat the superlattice period as the unit cell of a new crystal and take into account the actual atomic nature of the constituent materials. At the cost of considerable computational complexity , these *ab initio* methods provide a deeper insight in the electronic properties of the superlattice. However , for most cases of practical interest the envelope function approximation is found to work very well. A large part of the experimental work on the optical properties of semiconductor quantum wells has infact been analyzed on the basis of this intuitive theoretical method.

Throughout the discussion we will consider a general superlattice consisting of two different materials (A for the well and B for the barrier) of respective thicknesses L_z and L_b. The band structure of the layered heterostructure is calculated by matching the envelope functions at each interface. It is supposed that the wavefunctions of the constituent materials take the form [1]

$$\psi_{A,B}(\mathbf{r}) = F_j^{A,B}(\mathbf{r})U_j^{A,B}(\mathbf{r}) \tag{1}$$

In Eq.(1) $U_j(\mathbf{r})$ is the Bloch function at k=0 and $F_j(\mathbf{r}) = 1/\sqrt{S}\ \exp(i\ \mathbf{k}_\perp \cdot \mathbf{r}_\perp)\chi_j(z)$ where $\chi_j(z)$ is the slowly varying envelope function and S the quantum well surface. In doing so the following assumptions are explicitely made : i) the materials A and B are lattice matched and the interfaces are perfectly abrupt , i.e. $U^A(\mathbf{r}) = U^B(\mathbf{r})$; ii) only superlattice states close to the Γ point of the host materials are involved in the optical transitions. In the case of III-V heterostructures , only the usual Γ_6, Γ_7 and Γ_8 band edges contribute to the superlattice wavefunction ; iii) The electron dynamics of the superlattice is described by the envelope functions which are eigenstates of the superlattice Hamiltonian. All the details on the physical properties of the material are included in the band structure parameters of the constituent bulk crystals , namely , the band gaps , the effective masses and the Kane matrix elements.

In the realistic case of potential wells of finite depth, the calculation of the quantized subbands is performed by using a square well potential for uncoupled quantum wells (i.e. with wavefunctions totally confined in the well) , or a periodic Kronig-Penney potential for superlattice structures with coupled wells. In the former case , by applying the boundary conditions

$$\frac{1}{m_A}\frac{dF^A}{dz} = \frac{1}{m_B}\frac{dF^B}{dz} \tag{2}$$

which ensures the probability current conservation across the interfaces, the following dispersion relations can be found for even states in the well,

$$\frac{m_B}{m_A}K_A tan(\frac{K_A L_z}{2}) = K_B \tag{3}$$

and

$$\frac{m_B}{m_A} K_A cotan(\frac{K_A L_z}{2}) = -K_B \qquad (4)$$

for odd states confined in the well. For the case of unpenetrable (i.e. infinite) barriers the confinement energy of the electrons reduces to that calculated in the simplest particle-in-a-box model, with rectangular potential and effective electron mass m_A.

$$E_n = \frac{\hbar^2}{2m_A}(\frac{n}{L_z})^2 \qquad (5)$$

For superlattices with thin barriers, i.e. with strong coupling of the wavefunctions in different wells, the confined states merge into superlattice minibands. The use of a Kronig-Penney potential for these minibands results in the dispersion relation [2]

$$coskd = cos K_A L_z cosh K_B L_b + \frac{1}{2}[\frac{m_A K_B}{m_B K_A} - \frac{m_B K_A}{m_A K_B}]sin K_A L_z sinh K_B L_b \qquad (6)$$

In Eqs.(3-4) and (6)

$$K_A = \sqrt{\frac{2m_A E}{\hbar}} \qquad (7)$$

and

$$K_B = \sqrt{\frac{2m_B(V-E)}{\hbar}} \qquad (8)$$

where V is the potential barrier at the interface. The well width dependence of the GaAs gap calculated by means of eq.(5) is shown in Fig.2 for the case of quantum wells, wires and boxes with rectangular section and infinite barriers. The total confinement energy causes a strong blue shift of the crystal gap which varies in the spectral range between the near infrared and the green.

Large deviations from the behavior predicted by the independent particle theory is found due to the enhancement of the Coulomb interaction between electrons and holes in the heterostructures. For the ideal 2D case this leads to the formation of excitons with energy eigenvalues [3]

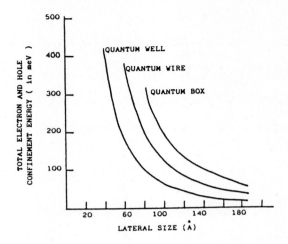

Fig.2 - Variation of the energy gap as a function of the physical size of a GaAs quantum well, quantum wire and quantum box (from eq .(5)).

$$E_n^{2D} = E_g - \frac{E_b}{(n - \frac{1}{2})^2} \qquad (9)$$

where E_g is the semiconductor gap and E_b is the three-dimensional exciton binding energy

$$E_b = \frac{e^4 \mu}{2\epsilon_o^2 h^2} \qquad (10)$$

μ being the exciton reduced mass and ϵ_o the static dielectric constant. In the limit of perfectly 2D system it results $E_b^{2D} = 4 E_b^{3D}$. Therefore the 2D exciton is more strongly bound and much more stable than the corresponding 3D quasi-particle. As a general trend, the reduction of the dimensionality results in the increase of the exciton binding energy due to the increased overlap of the electron and hole wavefunctions in the crystal. However, in real low dimensional semiconductors there is a strong dependence of the exciton binding energy on the actual size of the heterostructure, i.e. on the degree of confinement of the electron and hole wavefunctions. In particular for quantum wells, the relevant exciton parameters, namely binding energy, oscillator strength and Bohr radius , are found to monotonically vary between the extreme 2D and 3D values in the range 4 nm $< L_z <$ 50 nm. In Fig.3 we show the results of experiments and calculations on the well width dependence of the exciton binding energy in GaAs quantum wells . For large L_z values the quantity depicted in Fig.3 converge to the well known bulk values.

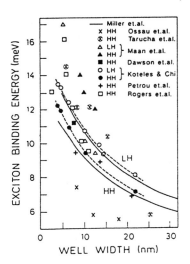

Fig.3 - *Comparison of the experimental and theoretical well width dependence of the exciton binding energy in GaAs quantum wells (after ref.[4]).*

The strong binding energy and the high oscillator strength of the quasi-two dimensional exciton result in the dominant excitonic character of the optical transitions in quantum wells. This is depicted in Fig.4 where we compare the absorption spectra of 3D and 2D semiconductors. The absorption coefficient for excitonic transitions can be expressed as

$$\alpha(\hbar\omega) \simeq |\langle U_e(\mathbf{r}) | \epsilon \cdot \mathbf{p} | U_h(\mathbf{r}) \rangle|^2 |\langle \chi_{n,h} | \chi_{n,e} \rangle|^2 |\Phi^l|^2 \, \delta_{K_\perp,0} \delta(E_{exc} - \hbar\omega) \ldots\ldots(10)$$

where the first term is the transition matrix element for the Bloch functions , $\chi_{n,e(h)}$ is the envelope function of the electron (hole) state of quantum number n, Φ^l is the exciton envelope function with angular momentum of quantum number l (usually s-states) , and the delta functions give the momentum and energy conservation in the transition (E_{exc} is the energy of the exciton state).

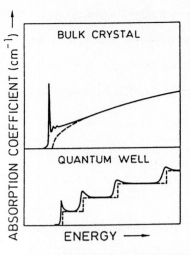

Fig.4 - Excitonic effects in bulk and two-dimensional semiconductors.

The following properties of the 2D exciton should be pointed out: i) the presence of the $|\Phi^l|^2$ term in Eq.(12) introduces a specific dependence of the excitonic absorption on the well width, as the exciton envelope function depends on L_z. ii) the narrow exciton peaks appearing in the absorption spectra reflect the momentum conservation of the excitonic transition. The k-conservation is relaxed in the presence of disorder, alloy fluctuation, interface roughness or carrier localization, resulting in broadened excitonic absorption structures; iii) In III-V semiconductor quantum wells the LO phonon interaction is reduced as compared to other strongly ionic materials (like the II-VI) semiconductors. Therefore, the increase of the exciton binding energy is sufficient to keep excitons bound at room temperature; iv) The exciton oscillator strength in quantum wells increases proportional to the shrinkage of the exciton wavefunction. This results in a strong enhancement of the excitonic features superimposed to the intraband continuum, as compared to the corresponding bulk material. In quantum wires and quantum dots similar considerations can be applied. The exciton binding energy is further increased with respect to the two-dimensional case, but the physical concepts remain the same.

2. Experimental studies of the electronic and excitonic states in quantum wells

Typical examples of photoluminescence excitation (PLE) and absorption spectra of $GaAs/Al_xGa_{1-x}As$, $Ga_{0.47}In_{0.53}As/Al_{0.48}In_{0.52}As$ multiple quantum wells are shown in Figs.5 and 6.

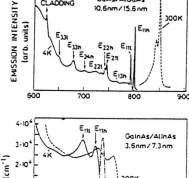

Fig.5 - Photoluminescence (PL) and Photoluminescence excitation (PLE) spectra of a GaAs/AlGaAs multiple quantum well.

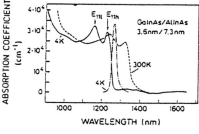

Fig.6 - Photoluminescence (PL) and Absorption (ABS) spectra of different InGaAs/AlInAs multiple quantum wells at 10 K and 300 K.

In these spectra the fundamental gap of the crystal is clearly blue-shifted with respect to the bulk material. Further, sharp exciton peaks associated with different quantized levels are superimposed to the step-like continuum of the two-dimensional density of states. The attribution of the fundamental excitonic transitions in the spectra of Figs.5 and 6 is made by calculating the confinement energies of the condution and valence subbands, as explained in Sect.1, and taking into account the exciton binding energies displayed in Fig.3. The experimental data can be fitted to the theoretically predicted transition energies , within the envelope function approximation , as a function of the offset value (V factor in eq.(8)). This has given the nowaday commonly accepted value for the conduction-to-valence band offset ratio (between 60:40 and 70:30), which is routinely used to fit the optical spectra of different III-V heterostructures.

In Figs.7 and 8 we also show for comparison the PLE spectra and the absorption spectra of an array of GaAs quantum wires and of a set of CdSeS quantum dots embebbed in a transparent glass matrix. .

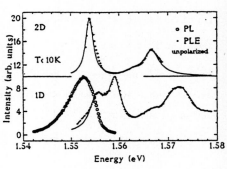

Fig.7 - *Photoluminescence excitation (PLE) and Photoluminescence (PL) spectra of GaAs/AlGaAs quantum wires (1D) and of the quantum well from which the wires were fabricated (2D)*

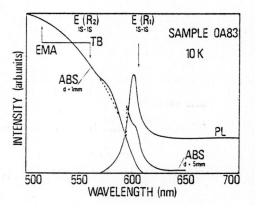

Fig.8 - *Absorption (ABS) and Photoluminescence (PL) spectra of CdSeS quantum dots at 10 K. The quantum dot size ranges between 2 and 5 nm.*

The quantum wires of Fig.7 have a lateral width of about 60 nm and a transverse width of about 10 nm. The lateral dimension of 60 nm is four times larger than the exiton Bohr radius of GaAs. This results in a weak quantum confinement of the electron and hole wavefunctions, i.e. a weak one-dimensional confinement, and in a clear quantization of the excitonic envelope function, i.e. in the quantization of the center of mass motion of the exciton. The observed sharp peaks in the PLE spectrum are thus due to the quantized states of the excitonic center of mass motion, which behave like standing waves in the wire. For narrower wires these effects are lost and the true electron and hole confinement is recovered, with the appearence of sharp features associated with the 1D density of states [6]. For the case of spherical quantum dots (Fig.8), the situation is a slightly different. Usually, these structures are prepared by complex chemical reactions and heat treatements. Their physical size depends on the temperature, duration and characteristic of the reaction [6]. This makes very difficult the nanometer control of the quantum dot size which are usually polidispersed over a rather broad size distribution. The analysis of the quantum dot size at the transmission electron miscroscope becomes thus crucial to determine the mean dot radius and the radius dispersion around the mean value. For radii of the order of the exciton Bohr radius (about 3 nm in CdSeS) the effective mass approximation can be applied to spherical quantum dots. The resulting quantized states fit satisfactorily the weak absorption structures observed in Fig.8 [7].

An unusual example of band gap engineering is given by the ultra-short-period GaAs/AlAs superlattices in which the well and the barrier slabs have typical thicknesses of the order of 1 nm. In this case due to the indirect nature of the AlAs barrier and to the strong confinement energy in the ultra-thin GaAs well, the X-point of the AlAs barrier lies at lower energy than the GaAs Γ-point. As a consequence, electrons at the AlAs X-point are confined in the barrier by the higher energy Γ-point of the GaAs layer. This results in a staggered band alignment (type II superlattice) in which the lowest conduction state of the superlattice is located in the AlAs barrier and has the symmetry properties of the AlAs X-point, while the top of the valence band is still located at the GaAs Γ-point. The effects of these peculiar electronic properties on the emission and absorption processes of the USPS are shown on the right-hand side of Fig.9. The luminescence arises from the lowest energy levels of the superlattice, i.e. from the X-Γ or type II transition, while the absorption has a considerable strength only above the direct type I ($\Gamma - \Gamma$) edge.

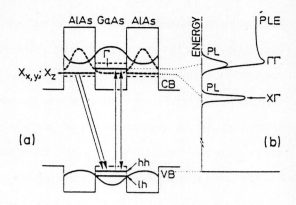

Fig.9 - Optical properties and electronic states in ultra-short-period superlattices.

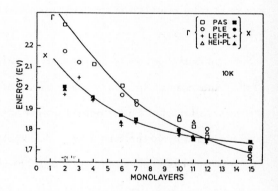

Fig.10 - Comparison between the measured (symbols) and the calculated type I ($\Gamma-\Gamma$) and type II ($\Gamma-X$) energy gaps of GaAs/AlAs USPS as a function of the number of monolayers constituting the well and the barrier. PAS indicates photoacoustic spectroscopy data.

Therefore, the comparison of PL and PLE provides complementary information on the type I and type II gaps in the USPS. Furthermore, we note that the type II transitions are indirect in real space, while the direct or indirect character in k-space is determined by the folding of the X-states onto the Γ-state, i.e. ultimately on the symmetry properties of the superlattice minizone [8]. In Fig.10, we show the comparison of a collection of photoluminescence, photoluminescence excitation and photoacoustic data [9] obtained from a set of symmetric $(GaAs)_m/(AlAs)_n$ USPS with the well width dependences of the type I and type II gaps calculated in the frame of the envelope function approximation and assuming a Kronig-Penney potential. The type II-type I cross-over occurs around 12 monolayers, in excellent agreement with the theory.

3. Applicative aspects

Most of the technological applications of low dimensional semiconductors are closely related to the possibility of getting tunable coherent emission and non-linear absorption. Both processes are usually obtained by generating a dense electron-hole plasma population by optical or electrical injection. Under high carrier density the Coulomb interaction is screened, excitons are ionized and a dense electron-hole plasma (EHP) forms. This strongly affects the optical properties of the crystal: many-body interactions in the dense EHP renormalize the electron-hole pair self-energy and change the gap of the crystal. All radiative recombination processes are governed by free-carrier optical transitions, and the conditions for optical amplification of the luminescence can be reached if population inversion is maintained by optical or electrical carrier injection. The impact of these effects on the energy bands and on the optical properties of the crystal is schematized in Fig.11.

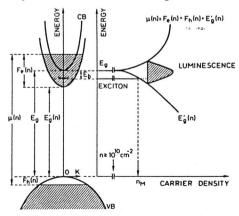

Fig.11 - *Energetics of the electron hole plasma phase in semiconductors (left), and carrier density dependence of the band gap and total quasi-Fermi level (right).*

The band gap of the crystal (E_g) shrinks proportional to the density of the

EHP ($E'_g(n)$), while electrons and holes fill the respective bands up to the Fermi levels ($F_e(n)$ and $F_h(n)$), following a quasi-equilibrium Fermi distribution. As a consequence, the EHP luminescence red-shifts and broadens with increasing carrier density, reflecting the shrinkage of the gap and the increase of the chemical potential $\mu(n)$. As schematically shown on the right hand side of Fig.11, over a certain critical density n_M the reduced band gap merges the exciton level, whose energy does not renormalize due to its charge neutrality. Under these conditions the exciton binding energy vanishes and excitons are no more stable in the EHP [10]. The critical density n_M for the exciton stability is found to be in the range $10^{10} - 10^{11} cm^{-2}$

The study of the electron hole plasma regime is of fundamental importance for the understanding of solid state lasers based on low dimensional semiconductors. The basic advantage of these systems is schematically depicted in Fig.12.

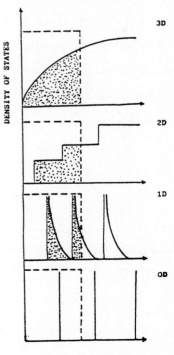

Fig.12 - Carrier density in low dimensional semiconductors. The integral $\int f(E)D(E)dE$ (the shaded areas) reduces with decreasing the crystal dimensionality.

With decreasing the dimensionality, the total carrier density $n = \int f(E)D(E)$ reduces, resulting in a reduction of the stimulated emission threshold (f(E) is the carrier distribution function and D(E) is the density of states). This is a direct consequence of the modification of the density of states in low dimensional systems. Furthermore, with increasing the chemical potential, i.e. the carrier density, higher energy quantized states can be progressively populated, giving complex band filling spectra in which recombination among different subbands can be identified. This is shown in Fig.13 for a GaAs quantum well and a GaAs quantum wire array as a function of the excitation intensity [11, 12].

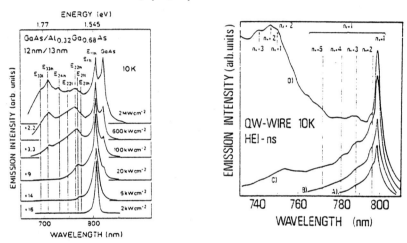

Fig.13 - Band filling photoluminescence spectra of GaAs quantum wells and quantum wires (left and right, respectively) measured at 10 K and under different excitation intensities. The intersubband transitions among quantized states are indicated.

It is now clear why so many efforts have been made in order to get stimulated emission and laser action from low dimensional systems, to exploit the advantageous band gap tunability and the low stimulated emission thresholds. Nowaday laser emission in quantum wells is almost continuously tunable in the range 600 nm - 1.6 μm, by suitably controlling the configurational parameters of different III-V semiconductor heterostructures. The situation is by far less advanced for the case of wires and dots, due to great technological difficulties in the fabrication of high quality heterostructures. Details on the physics of the quantum well lasers can be found in ref.[13]. A typical structure for a quantum well laser is shown in Fig.14. This is called graded-index-separate-confinement heterostructure, and is realized by a variable compositional profile of the external barriers which provides a parabolic or triangular potential profile around the active quantum well. This enhances the carrier trapping into the emitting state, which are first collected into the wide quantum well region and then relax at the lowest emitting state, and also provides a favourable refractive index profile for light waveguiding.

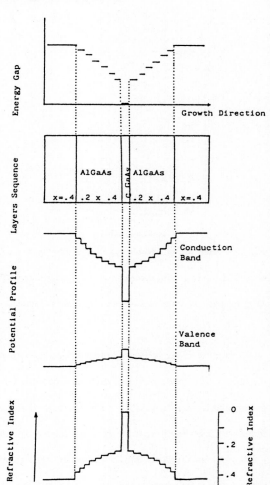

Fig.14 - *Compositional profile, energy band scheme and refractive index profile of a GaAs/AlGaAs graded-index-separate-confinement heterostructure laser.*

References

[1] G.Bastard, Phys. Rev.B24,5693 (1981)

G.Bastard, Phys. Rev. B25,7584 (1982)

G.Bastard, *"Wave Mechanics Applied to Semiconductor Heterostructures"*, Les Editions Physique, Les Ulis (France), 1988

[2] Huang-Sik Cho and P.R.Prucnal, Phys. Rev. B36,3237 (1987)

[3] M.Shinada and M.Sugano, J. Phys. Soc. Jpn. 21, 1936 (1966)

[4] E.S.Koteles and J.C.Chi, Phys. Rev. B37, 6332 (1987)

[5] H.Lage, D.Heitmann, R.Cingolani, P.Grambow and K.Ploog, Phys. Rev. B44, 6550 (1991)

[6] L.E.Brus, J.Chem.Phys. 80, 4403 (1984)

[7] R.Cingolani, C.Moro, D.Manno, M.Striccoli, C.DeBlasi, G.C. Righini, and M.Ferrara, J. Appl. Phys. 70, 6898 (1991)

[8] Y.T.Lu and L.J.Sham , Phys. Rev. B40 , 5567 (1989)

[9] R.Cingolani , L.Baldassarre , M.Ferrara , M.Lugara' , and K.Ploog , Phys. Rev. B40 , 6101 (1989)

R.Cingolani , K.Ploog , L.Baldassarre , M.Ferrara and M.Lugara' , Appl. Phys. A50 , 189 (1989)

[10] S.Schmitt-Rink and C.Ell , J. Lumin. 30 , 585 (1985)

[11] R.Cingolani , K.Ploog , A.Cingolani , C.Moro , and M.Ferrara , Phys. Rev. B42 , 2893 (1990)

[12] R.Cingolani, H.Lage, L.Tapfer, D.Heitmann, H.Kalt, K.Ploog, Phys. Rev. Lett. 67, 691 (1991)

[13] C.Weisbuch and J.Nagle , Phys. Scr. T19 , 209 (1987)

Electronic and optical properties of low-dimensional semiconductor structures

Joachim Wagner
Fraunhofer-Institut für Angewandte Festkörperphysik,
Tullastrasse 72, D-7800 Freiburg, Federal Republic of Germany

Abstract. This paper reviews the use of photoluminescence and photoluminescence excitation spectroscopy for the study of low-dimensional electron and hole gases in modulation- and δ-doped semiconductor heterostructures. This type of spectroscopy provides information on the subband structure and band filling in those structures. In addition, insight is gained in electron-hole many-body interactions, such as the so-called Fermi edge singularity, and in the electron-hole interaction strength. The latter information can be extracted from, e.g., the optical measurement of electron spin relaxation kinetics. Selected examples, mainly based on $GaAs/Al_xGa_{1-x}As$ heterostructures, will be discussed.

1. Introduction

There is considerable current interest in the physics of low-dimensional electron and hole gases in semiconductors. The work-horse for magneto-transport experiments on two-dimensional electron gases (2DEG) has been in the past, as far as III-V semiconductors are concerned, the modulation-doped $GaAs/Al_xGa_{1-x}As$ heterojunction [1]. Besides transport measurements, far-infrared absorption spectroscopy [2] and Raman spectroscopy [3] have also been used to study elementary excitations of the 2DEG at such heterojunctions. Photoluminescence (PL) and photoluminescence excitation (PLE) spectroscopy, in contrast, which have proven to be powerful techniques for the investigation of undoped quantum well structures [4], have been applied only recently to modulation doped heterojunctions [5-7]. Both in these heterojunctions and in so-called δ (or planar) doped structures, where a sheet of dopant atoms generates a space-charge induced potential well confining the 2DEG, the potential confining the *majority* carriers is repulsive for the photogenerated *minority* carriers. This repulsion results in a small electron-hole wavefunction overlap and consequently in a low, or even vanishing, PL intensity.

In the present paper we shall focus on PL and PLE spectroscopy of 2DEG and two-dimensional hole gas (2DHG) systems, both in modulation-doped and δ-doped GaAs/Al$_x$Ga$_{1-x}$As heterostructures, and show how sufficiant confinement of the optically generated minority carriers can be achieved to enhance the radiative recombination efficiency. The information on electron subband structure and band filling gained from these experiments will be compared with self-consistent potential and subband calculations. Further, the effect of many-body interactions on the optical absorption and emission spectra of low-dimensional electron systems, the so-called Fermi edge singularity (FES) [8-10], will be discussed. Electron spin relaxation kinetics can be studied by time-resolved PL spectroscopy using circularly polarized light [11] and it will be shown that information on the electron-hole interaction strength can be extracted from such experiments.

2. Spatially indirect photoluminescence in modulation- and δ-doped heterostructures

2.1 Modulation-doped GaAs/Al$_x$Ga$_{1-x}$As heterojunctions

A schematic energy band diagram of a n-type modulation-doped GaAs/Al$_x$Ga$_{1-x}$As heterojunction is shown in Fig. 1. From this diagram it is evident that the potential well at the heterointerface generated by the ionized donors in the Al$_x$Ga$_{1-x}$As supply layer leads to a repulsion of minority carriers (holes) away from the 2DEG located at that interface. Photoluminescence arising from the radiative recombination of photogenerated holes with electrons from the 2DEG is therefore expected to be strongly suppressed because of the spatial separation of electrons and holes leading to a poor electron-hole wavefunction overlap. However, as it has been shown by Chang et al. [5], even under those unfavorable conditions photoluminescence from the 2DEG can be resolved if optical excitation is performed with light of appropriate photon energy. Fig. 2 displays low-temperature (4 K) PL spectra of a Si modulation-doped GaAs/Al$_{0.3}$Ga$_{0.7}$As heterojunction excited in the IR (1.648 eV) (a) and in the UV (3.535 eV) (b), respectively [5]. For IR excitation the Al$_{0.3}$Ga$_{0.7}$As supply layer is transparent to the incident light which gets absorbed in the GaAs buffer layer. Consequently, the PL spectrum is dominated by donor-bound exciton

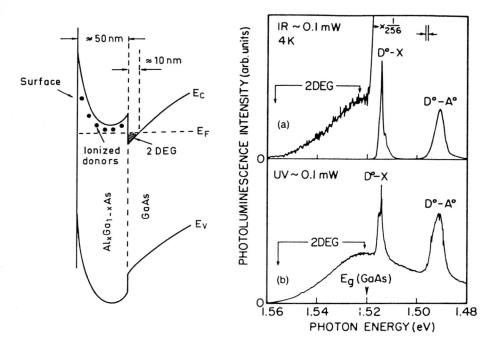

Fig. 1 (left) Schematic real-space energy band diagram of a modulation-doped GaAs/Al$_{0.35}$Ga$_{0.65}$As heterojunction.

Fig. 2 (right) Low-temperature PL spectra of a n-modulation-doped GaAs/Al$_{0.3}$Ga$_{0.7}$As heterojunction excited (a) in the IR (1.648 eV) and (b) in the UV (3.535 eV), respectively (Ref. 5).

recombination (D^0-X) and donor-to-acceptor pair recombination (D^0-A^0) originating from that layer. Under UV excitation, where the incident light is entirely absorbed in the 65 nm thick Al$_{0.3}$Ga$_{0.7}$As supply layer, an additional broad emission is clearly resolved extending to energies well above the GaAs band gap energy. For IR excitation the relative intensity of this emission is drastically reduced. Based on this finding and on the observed changes in the spectrum upon applying an external voltage via a gate electrode, this emission has been assigned to radiative recombination involving the 2DEG at the heterojunction [5]. The fact that the 2DEG luminescence extends to energies above the band gap energy, the so-called Burstein-Moss shift [12], is due to a filling of the conduction band states at the degenerate electron gas with an areal electron density of 2.6x10^{11} cm^{-2} [5].

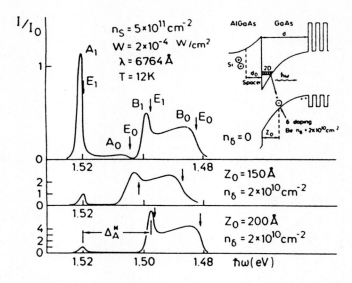

Fig. 3 Low-temperature PL spectra of radiative recombination of a 2DEG with photoexcited holes in a n-modulation-doped GaAs/Al$_x$Ga$_{1-x}$As heterojunction with a Be δ-doping layer inserted. The corresponding real-space energy band diagram is shown in the inset (Ref. 6).

To improve the intensity of 2DEG related PL in modulation-doped heterostructures, Kukushkin et al. [6,7] introduced a sample structure where a GaAs/AlAs short period superlattice was placed typically 50 nm below the GaAs/Al$_x$Ga$_{1-x}$As heterojunction to provide some confinement for the photogenerated holes [6]. A further improvement of the minority carrier confinement was achieved introducing a δ-doping layer of Be acceptors located some 20 nm underneath the heterointerface [7]. Typical PL spectra of such sample structures with a 2DEG concentration of 5×10^{11} cm^{-2} are displayed in Fig. 3 together with a real space energy band diagram of that structure [7]. The topmost spectrum recorded from a sample without a Be δ-doping layer shows as the dominant feature 2DEG emission involving free holes (labelled A_0 and A_1) together with a somewhat weaker emission involving holes bound to residual acceptors (B lines). E_0 and E_1 mark the onsets of corresponding transitions involving the first and second electron subband, respectively. With a Be δ-doping layer of an areal density of 2×10^{10} cm^{-2} placed 15 nm or 20 nm underneath the heterojunction (middle and lower specrum in Fig. 3, respectively) recombination involving acceptor bound holes gains in relative intensity (B lines) making this kind of luminescence spectra a handy tool for the study of 2DEG properties in, e.g., an applied external magnetic field [7]. Also the time dependence of the hole recombination in n-modulation-doped GaAs/Al$_x$Ga$_{1-x}$As heterostructures has been studied in some detail [13].

Here it should be noted that PL spectroscopy of the 2DEG in modulation-doped structures can now be performed at high magnetic fields even in the sub- and milli-Kelvin temperature range. This allows the optical investigations of, e.g., the integer and fractional quantum Hall effects [14]. Even inelastic light scattering (Raman scattering) experiments have been reported recently in the above temperature range revealing a collapse of the intersubband transition energy in the regime of the fractional quantum Hall effect [15]. These Raman experiments became possible exploiting extreme resonance conditions to enhance the light scattering signal. Usually Raman experiments require incident light intensities significantly higher than those required for PL spectroscopy because of the weak signal intensity, making experiments in the milli-Kelvin range very difficult.

2.2 Delta-doping layers in GaAs

An isolated doping spike of, e.g., Si or Be in GaAs forms a quasi-two-dimensional electronic system [16]. The electrons and holes are confined in a space-charge-induced potential well generated by the sheet of dopant atoms. In this well a 2DEG or a 2DHG is present where usually several subbands are occupied [16,17]. As an example Fig. 4 shows the self-consistently calculated potential profile for electrons in n-type δ-doped GaAs. The doping spike with a donor concentration of 8×10^{12} cm^{-2} was assumed at a nominal depth of 30 nm underneath the sample surface allowing for a segregation of the dopant atoms by up to 20 nm towards the surface [18]. It is known that nominally δ-doped layers may often show a considerable spread of the dopant atoms, in particular for high dopant sheet densities and sample growth at substrate temperatures > 500 °C, due to dopant diffusion and segregation [18,19,20]. Fig. 4 displays further the resulting electron subband energies and the probability densities. Because of Fermi level pinning at the surface due to surface electron traps the potential well shows a pronounced asymmetry and the free electron concentration is only 3.8×10^{12} cm^{-2}.

The energy spacing between the various subbands formed can be studied by direct intersubband absorption spectroscopy [21] as well as by inelastic light scattering [18,22]. In Fig. 5 the intersubband Raman spectrum of the above

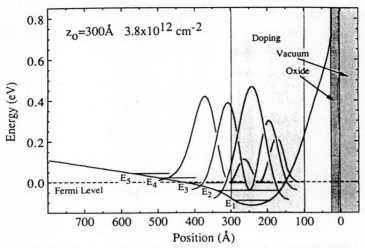

Fig. 4 Self-consistent electron potential, probability densities, and energy levels for n-type δ-doped GaAs. The doping spike of 8×10^{12} donors/cm^2 was placed at a nominal depth of $z_0 = 30$ nm. The dopant atoms are spread over 20 nm towards the surface. The resulting free carrier concentration is 3.8×10^{12} cm^{-2} (Ref. 18).

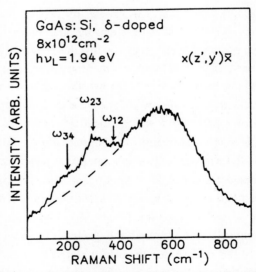

Fig. 5 Low-temperature (15 K) Raman spectrum of Si δ-doped GaAs with the doping spike placed at a nominal depth of $z_0 = 30$ nm. Sample growth was performed by molecular beam epitaxy at a substrate temperature of 580 °C. The polarized spectrum [$x(z',y')\bar{x}$] has been excited at 1.94 eV in resonance with the $E_0 + \Delta_0$ band gap of GaAs. ω_{12}, ω_{23}, and ω_{34} denote calculated spin-density excitation energies (see Fig. 4) (Ref. 18).

described sample structure is plotted. The spectrum shows spin-density intersubband transitions, labelled ω_{12}, ω_{23}, and ω_{34}, superimposed on a broad background from radiative recombination across the $E_0+\Delta_0$ band gap of GaAs [18]. The arrows mark calculated spin density excitation energies, based on the calculated potential profile shown in Fig. 4, which are in good agreement with the measured energies for transitions between the first and the second (ω_{12}), second and third (ω_{23}), and third and fourth (ω_{34}) subbands, respectively [18].

Similar to the above discussed case of modulation-doped heterostructures, for single δ-doping spikes PL spectroscopy is also complicated by the repulsive interaction of the photogenerated minority carriers with the potential confining the 2DEG or 2DHG. In n-type δ-doped GaAs this repulsion leads to a too large spatial separation of the carriers to observe radiative recombination from the 2DEG [23,24]. Placing the doping spike within a few ten nm below the surface some confinement of the photogenerated holes at the surface side could be achieved but the observed PL intensity was still weak because of competing surface recombination [25]. Replacing the GaAs surface by an $Al_xGa_{1-x}As$/GaAs heterointerface the 2DEG emission intensity was found to be strongly enhanced [25], and placing a second $Al_xGa_{1-x}As$ barrier at the substrate side of the doping spike a further drastic increase in 2DEG emission intensity could be achieved [26].

In Fig. 6 the self-consistent potentials for the 2DEG and for holes is plotted for such a δ-doped GaAs/$Al_xGa_{1-x}As$ double-heterostructure [27]. The n-type doping spike of 8×10^{12} donors/cm^2 is placed in the center of a 60 nm wide GaAs layer sandwiched between a 20 nm thick $Al_{0.33}Ga_{0.67}As$ barrier on the substrate side (left) and a corresponding 10 nm wide barrier at the surface side (right). The doping layer is assumed to be spread out towards the surface resulting in a $\Delta z = 10$ nm wide homogeneously doped layer. Electron and hole subband energies and probability densities are also indicated. This potential diagram shows that there is a clear confinement of photogenerated holes at either side of the potential well confining the 2DEG which leads to a finite electron-hole wavefunction overlap. On the other hand the heterointerfaces are sufficiently remote such that the 2DEG is confined entirely by the space-charge-induced potential well and not by the conduction band discontinuities.

Low-temperature (6 K) PL spectra of such a δ-doped $Al_xGa_{1-x}As$/GaAs:Si/$Al_xGa_{1-x}As$ double-heterostructure are plotted in Fig. 7 for various excitation power densities [27]. Strong emission is observed from the 2DEG involving free

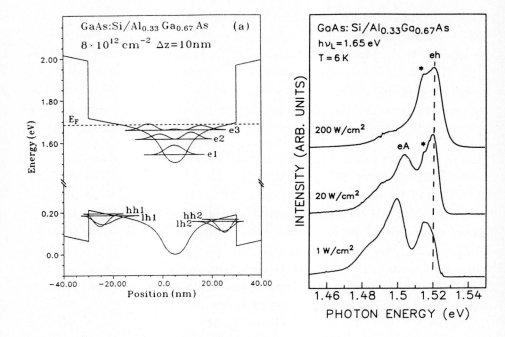

Fig. 6 (left) Self-consistent potential in a δ-doped GaAs:Si/Al$_{0.33}$Ga$_{0.67}$As double-heterostructure. The surface is on the right at 42 nm. A Si concentration of 8×10^{12} cm^{-2} is taken with a 10 nm dopant spread towards the surface. Energies and probability densities are shown for electron subbands ei and for the first heavy- (hh1) and light-hole (lh1) subbands at $k_{//}=0$ (Ref. 27).

Fig. 7 (right) Excitation intensity dependent PL spectra of a δ-doped GaAs:Si/Al$_{0.33}$Ga$_{0.67}$As double-heterostructure with a dopant concentration of 8×10^{12} cm^{-2}. The spectra were recorded at 6 K with excitation at 1.65 eV. *eh* and *eA* denote band-to-band and band-to-acceptor emission from the 2DEG. The asterisk marks emission from the undoped GaAs buffer layer (Ref. 27).

holes (eh) and holes bound to residual acceptors (eA). With increasing power density the eA emission saturates leaving the eh recombination as the dominant feature. Bound exciton emission from the underlying undoped GaAs buffer layer is comparatively weak and just resolved as shoulder on the low energy side of the eh emission band (marked by asterisks). With increasing excitation intensity the 2DEG emission spectrum shifts to higher energies. This energy shift is characteristic for spatially *indirect* transitions. In the present sample structure, where electrons and holes are spatially separated, the strong spatial modulation

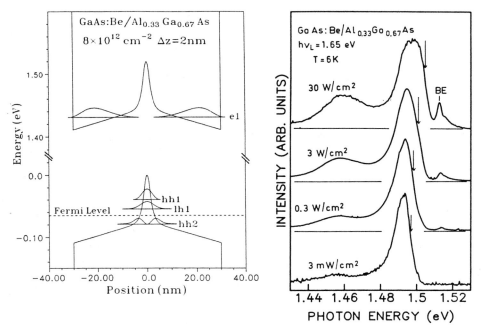

Fig. 8 (left) Self-consistent potential profile for the 2DHG in a δ-doped GaAs:Be/Al$_{0.33}$Ga$_{0.67}$As double-heterostructure. An 8x10^{12} cm^{-2} acceptor concentration is taken with a 2 nm dopant spread. Subband energies and probability densities at $k_{//} = 0$ are also shown. 'hh' and 'lh' denote heavy-hole and light-hole subbands, respectively; 'e1' denotes the first electron subband (Ref. 46).

Fig. 9 (right) Excitation intensity dependent PL spectra of a δ-doped GaAs:Be/Al$_{0.33}$Ga$_{0.67}$As double-heterostructure with a dopant concentration of 3x10^{13} cm^{-2}. The spectra were recorded at 6 K with excitation at 1.65 eV. BE denotes bound exciton emission from the undoped GaAs buffer layer (Ref. 29).

of the band edge energies due to space charge effects leads to a lowering of the emission energy as compared to spatially *direct* transitions. With increasing excitation intensity photogenerated carriers cause an increasing screening of the space-charge potential resulting in a high-energy shift of the spatially *indirect* emission.

In p-type δ-doped GaAs the repulsion of the minority carriers by the hole confining potential seems to have a less detrimental effect on the 2DHG luminescence efficiency because radiative recombination from the 2DHG has been reported from plain Be δ-doped GaAs layers [28]. However, also in this case placing the doping spike in the center of an $Al_xGa_{1-x}As/GaAs/Al_xGa_{1-x}As$ double-heterostructure leads to a significant enhancement of the 2DHG emission intensity [29,30]. Fig. 8 shows the self-consistent potential for electrons and holes for such a p-type δ-doped $Al_xGa_{1-x}As/GaAs:Be/Al_xGa_{1-x}As$ double-heterostructure. The structure parameters and the doping density of 8×10^{12} cm^{-2} are identical to that of Fig. 6 except for the reduced spread of the doping layer of $\Delta z = 2$ nm [30]. Compared to n-type δ-doping the width of the space-charge induced potential well is significantly reduced because of the larger effective hole mass [30].

PL spectra of a double-heterostructure with a higher p-type doping concentration of 3×10^{13} cm^{-2} are plotted in Fig. 9 for various excitation energies. Two 2DHG emission bands are observed resulting from recombination involving the first and the second heavy hole subband, respectively [30]. Weak bound exciton emission (BE) from the GaAs buffer layer is also observed. There are again changes in the 2DHG emission spectrum with increasing excitation intensity due to the above mentioned screening effects and due to band-filling by photogenerated carriers [29].

3. Fermi edge singularity as an example for many-body effects in the optical spectra of low-dimensional electron and hole gases

In two-dimensional systems with low carrier concentrations, such as undoped quantum wells under low optical excitation, optical interband absorption and emission are governed by excitonic effects. These effects lead to an excitonic enhancement of the band-to-band continuum absorption for photon energies close to the band edge energy as well as to the occurrence of sharp exciton absorption and emission lines just below the band gap energy. If the carrier concentration is increased, either by the optical generation of electron-hole pairs or by electrical injection of carriers in, e.g., a forward biased p-n-junction, exchange effects and the screening of the Coulomb interaction destabilize the exciton and an electron-hole plasma is formed [8]. Similarly, if large concentrations of either electrons or holes are introduced by doping, such as in modulation-doped heterojunctions and quantum wells, excitons get screened [10]. However, even though electron-hole bound states, namely excitons, do not exist at high carrier densities, multiple electron-hole scattering and Coulomb interactions are still very pronounced in particular in two-dimensional systems [8,9]. This leads to an

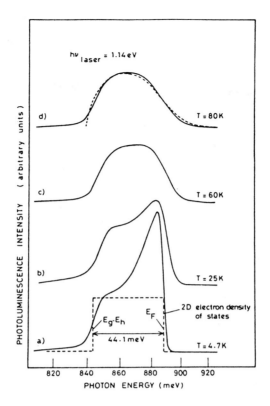

Fig. 10 PL spectra recorded at 4.7, 25, 60, and 80 K from a n-modulation-doped $In_xGa_{1-x}As/InP$ quantum well with a free electron concentration of 9.1×10^{11} cm^{-2}. The spectrum at 4.7 K in (a) shows a strong enhancement towards the Fermi energy E_F. The theoretical spectrum corresponding to the ideal two-dimensional density of states is superimposed to emphasize the Coulomb enhancement (FES) effects. With increasing temperature the enhancement effects are reduced, and are absent at 80 K where a good fit is obtained assuming no enhancement [the dashed curve in (d)] (Ref. 9).

enhanced probability for optical interband transitions involving electron states close to the Fermi energy E_F. For electrons well below E_F this effect is suppressed because of the exclusion principle.

The first direct experimental manifestation of this effect has been reported by Skolnick et al. [9]. These authors observed a strong enhancement of the photoluminescence intensity towards the electron Fermi energy in n-type modulation-doped $In_xGa_{1-x}As/InP$ quantum wells [9]. Fig. 10(a) shows a low-temperature (4.7 K) photoluminescence spectrum of such a sample structure. The observed shape of that luminescence spectrum deviates significantly from the expected square-like shape (indicated by the dashed lines) due to the constant two-dimensional density of states. The strong enhancement of the emission intensity [the so-called Fermi edge singularity (FES)] at energies close to E_F is clearly resolved. With increasing temperature the Fermi edge enhancement weakens and the FES disappears completely for temperatures exceeding 25 K [see Fig. 10(b-d)]. This decrease with temperature, which is a characteristic finger print for a FES, is due to the increasing energy spread of the electrons around E_F with increasing temperature.

To observe a FES in luminescence, electrons with energies close to E_F, and consequently a wavevector close to the Fermi wavevector k_F, have to recombine with photogenerated holes relaxed to the top of the valence band. These holes have a wavevector of essentially zero. If only momentum conserving transitions were allowed, this recombination would be forbidden and therefore, the FES should be strongly suppressed. In the case of the above discussed $In_xGa_{1-x}As$ quantum wells the observation of a strong FES has been explained by a localization, and thus a spread in k-space, of the holes [9].

Another mechanism for the observation of a FES in PL, despite the mismatch in electron and hole wavevector, has been suggested by Chen et al. [31,32]. They studied modulation-doped quantum wells with the doping level adjusted such that the electron Fermi level lies just below the second unoccupied electron subband. Excitation intensity dependent PL spectra of such a 20 nm wide n-type modulation doped GaAs quantum well are shown in Fig. 11 [32]. At low excitation intensities there are two distinct peaks in the high-energy portion of the PL spectrum. The one at higher energies arises from exciton recombination involving the second unoccuppied subband. The maximum at somewhat lower energies is attributed to a FES. With increasing optical excitation intensity the electron Fermi energy raises getting closer to the second electron subband and the FES increases in intensity. This observation of a FES is unique to samples with E_F close to the second electron subband. For samples with higher or lower electron concentrations, respectively, no Fermi edge enhancement was observed (see inset in Fig. 11). Further evidence, that the observation of a FES is directly related to the fact that the Fermi level and the unoccupied subband are almost degenerate, was obtained from magneto-PL measurements [32]. In Fig. 12 the intensity of the FES and the strength of emission from the lowest electron subband are plotted as a function of the magnetic field applied along the direction

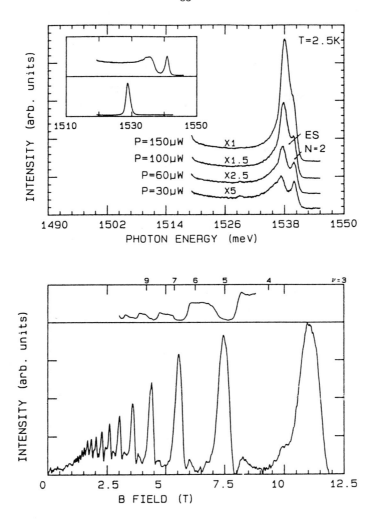

Fig. 11 (top) PL spectra of a n-modulation-doped GaAs quantum well in the region of the n=2 conduction to n=1 heavy-hole transition at 2.5 K, showing the Fermi edge singularity (ES) and its dependence on the optical excitation intensity, as well as the N=2 exciton. The inset shows spectra from two other samples with different carrier concentrations (Ref. 32).

Fig. 12 (bottom) Amplitude of the Fermi edge singularity as a function of magnetic field at 2.5 K reaching into the high-field regime (lower panel). The upper panel shows the variations in amplitude of the much weaker luminescence from electrons in the lower Landau levels of the n=1 subband (Ref. 32).

of quantization [32]. A strong enhancement of the FES intensity and very pronounced amplitude variations are observed with increasing magnetic field. These amplitude variations coincide with a variation of the energy separation between E_F and the lowest Landau level of the unoccupied electron subband [31,32]. This periodic change in energy separartion with increasing magnetic field is caused by the increasing number of depopulated Landau levels.

These observations have been interpreted as a hybridization of the occupied states at the Fermi edge with virtual excitons involving the unoccupied subband [31-34]. This hybridization can cause a large enhancement of the optical matrix element. The whole process becomes analogous to an edge singularity occuring at zero wavevector and therefore the need for localization of the minority carriers (holes) is relaxed [33,34].

Multiple electron-hole scattering at the Fermi edge leads also to an enhancement of the optical interband absorption for energies close to E_F [8,10]. In this case constraints imposed by wavevector conservation have less effect on the experimental observation of the FES because in absorption electron-hole pairs can be generated at any wavevector including k_F. As an illustration Fig. 13 shows low-temperature PL and PL excitation (PLE) spectra of a n-type δ-doped GaAs/Al$_x$Ga$_{1-x}$As double-heterostructure [26] (see Figs. 6 and 7). The PLE spectrum recorded with the detection set to the band-to-band (eh) recombination line, which essentially reflects the interband absorption spectrum, shows a well resolved enhancement in absorption just above the Fermi edge. This FES is not observed, in contrast, when the detection is set to the band-to-acceptor (eA) recombination line. In this case the PLE spectrum just reflects the intrinsic absorption of the 60 nm wide GaAs layer in which the doping spike is embedded.

So far the discussion of many body-effects at the Fermi edge has been limited to n-type doped systems. Recently a FES has also been observed in the PL spectrum of a high-density 2DHG at a p-type δ-doping layer in GaAs [29]. Fig. 14 shows temperature-dependent PL spectra of such a structure (see also Figs. 8 and 9). For optical excitation at 3.00 eV high above the band gap energy of the Al$_{0.33}$Ga$_{0.67}$As barriers a well resolved FES is observed in the luminescence spectrum for temperatures $\leq$ 15 K. In contrast to the 2DEG systems discussed before, here we deal with electrons as the minority carriers occupying states with zero wavevector. Unlike the first case of a FES in the emission spectrum of a 2DEG, where hole localization was assumed, in the present case of a 2DHG the electrons are unlikely to be localized because of their

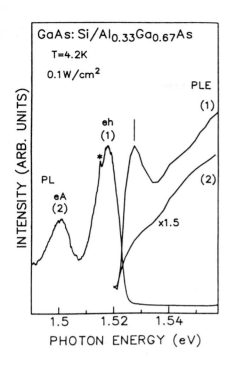
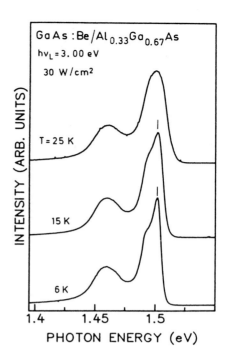

Fig. 13 (left) Low temperature PL (left) and PLE (right) spectra of a δ-doped GaAs:Si/Al$_{0.33}$Ga$_{0.67}$As double-heterostructure with a dopant concentration of 8×10^{12} cm^{-2}. For recording the PLE spectra detection was set either to the band-to-band (eh) (1) or to the band-to-acceptor (eA) recombination (2). The PL spectrum was excited at 1.559 eV. Excitation intensity was 0.1 W/cm^2. The Fermi edge enhancement in the PLE spectrum (1) is marked by a vertical line. The asterisk marks emission from the undoped GaAs buffer layer (Ref. 26).

Fig. 14 (right) Temperature-dependent PL spectra of a δ-doped GaAs:Be/Al$_{0.33}$Ga$_{0.67}$As double-heterostructure with a dopant concentration of 3×10^{13} cm^{-2} excited at 3.00 eV. The vertical lines mark the enhancement in luminescence intensity at the Fermi edge (Ref. 29).

much smaller mass. Instead one has to assume that the observation of a FES in the PL spectrum of the 2DHG is brought about via the hybridization between states at the Fermi edge and an unoccupied hole subband lying close to the hole Fermi energy [30]. That such unoccupied subbands are indeed expected to lie close to the Fermi level is seen from the self-consistent hole potential shown in Fig. 15 [30]. Two unoccupied subbands, namely the third heavy-hole (hh3) and

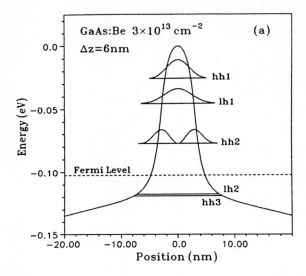

Fig. 15 Self-consistent potential profile for the 2DHG in a δ-doped GaAs:Be/Al$_{0.33}$Ga$_{0.67}$As double-heterostructure. An 3x10^{13} cm^{-2} acceptor concentration is taken with a 6 nm dopant spread. Subband energies and probability densities at $k_{//}= 0$ are also shown. 'hh' and 'lh' denote heavy-hole and light-hole subbands, respectively (Ref. 30).

the second light-hole (lh2) band, lie only 10 to 20 meV away from the Fermi level. The existence and energy position of these subbands has been verified experimentally by temperature dependent PL measurements where for temperatures ≥ 40 K the hh3 and lh2 subbands become thermally occupied and show strong luminescence [30].

So far we have discussed the FES in *two-dimensional* electron or hole gases. An interesting point to be addressed now is the possible strength of a FES in *one-dimensional* (1D) systems. On one hand theoretical calculations indicate that the Sommerfeld factor, which gives the excitonic enhancement of the absorption at the band edge over that of free electron-hole pairs above the band edge, is less than unity in the one-dimensional case [35] making the occurence of a strong FES less likely. On the other hand the singularity of the 1D density of states at the subband edges tends to enhance such effects. Calleja et al. [36] studied this question experimentally by looking at the 1D electron gas in modulation doped GaAs quantum wires. They found in structures which were in the 1D quantum limit, i.e. only the lowest 1D electron subband is occupied, a strong FES both in

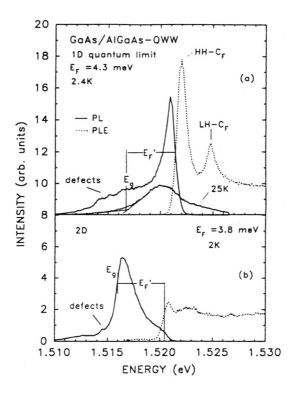

Fig. 16 Low-temperature PL (solid lines) and PLE (dashed lines) spectra (a) of quantum wires with a geometrical width of 100 nm fabricated from a 25 nm wide n-modulation-doped quantum well and (b) of an unpatterned sample with comparable Fermi energy E_F. $E_F' = E_F(1+m_e/m_h)$, where m_e and m_h are the electron and hole effective masses (Ref. 36).

the PL and the PLE spectrum (see Fig. 16). In the unstructured 2DEG reference sample the FES was much weaker in the PLE spectrum and completely lacking in the 2DEG PL spectrum [36]. However, the authors did not exclude that also a hybridization between the states at the Fermi edge and nearby lying unoccupied subbands may contribute to the strong FES in the 1D case. Thus the possible strength of a FES in 1D structures, as well as the maximum temperature to which this effect persists, remain an open question calling for further experimental and theoretical work.

4. Electron spin relaxation kinetics in p-type doped heterostructures: a sensitive probe for the electron-hole wavefunction overlap

Polarization dependent measurements of optical interband transitions have been proven to be a powerful method to study the dynamics of carrier and spin relaxation in semiconductors [11,37-40]. In p-type doped semiconductors, it is possible to study spin-polarized conduction electrons and their spin relaxation kinetics by luminescence polarization spectroscopy [11,37-43]. Low-temperature spin relaxation times in p-type bulk doped GaAs are $\leq$ 200 ps for acceptor concentrations $\geq$ 10^{19} cm^{-3} (Ref. 43). Based on measured hole density and temperature dependences it was suggested that electron-hole scattering with a simultaneous exchange interaction (the so-called Bir-Aronov-Pikus (BAP) mechanism [44]) dominates electron spin relaxation at low temperatures and high doping levels. In contrast, the Elliott-Yafet (EY) [41,42] and D'yakonov-Perel' (DP) mechanisms [45] should determine the relaxation rates at high temperatures and/or low hole concentrations [41-43]. An enhancement of the spin relaxation rate by a factor of three to four, as compared to bulk material with comparable hole densities, has been observed in p-modulation-doped quantum wells [11]. This has been attributed to the enhanced electron-hole wavefunction overlap in quantum wells and has been taken as additional evidence for the BAP mechanism being the dominant one at low temperatures [11]. In modulation doped heterojunctions and δ-doped double-heterostructures, in contrast, the electron-hole wavefunction overlap is drastically reduced. Therefore such structures are expected to show a drastic enhancement of the electron spin relaxation time.

In the following we shall discuss time-resolved circularly polarized PL experiments on a δ-doped GaAs:Be/Al$_x$Ga$_{1-x}$As double-heterostructure, in which the photogenerated electrons are spatially separated from the holes, while still maintaining sufficient wavefunction overlap for efficient radiative recombination to probe the spin polarization of the electrons [46] (see Sec. 2.2 and Fig. 8).

Fig. 17 shows low-temperature cw PL spectra of such a p-type δ-doped double-heterostructure with a dopant concentration of 8×10^{12} cm^{-2}. Polarized (I^+) and depolarized (I^-) luminescence spectra were recorded with the detected light having the same and the opposite circular polarization of the pump light, respectively. In the total intensity spectrum ($I^+ + I^-$) plotted in Fig. 17(a) the emission from the two-dimensional hole gas (2DHG) appears as a broad band centered around 1.47 eV. The peaks marked by asterisks at 1.513 and 1.493 eV arise from, respectively, bound-exciton and donor-to-acceptor pair recombination in the GaAs buffer layer. The difference spectrum (I^+-I^-) [Fig. 17(b)] shows a maximum at 1.46 eV and a minimum at 1.48 eV. Based on the polarization selection rules [37,41-44] these extrema are assigned to recombination involving the hh1 and the lh1 subband, respectively. The spacing of the calculated subband energies, marked in Fig. 17(b), is in good agreement with the difference between the experimental transition energies. Fig. 17(c) displays the degree of circular polarization $P=(I^+-I^-)/(I^++I^-)$ plotted versus the emitted photon energy. For the hh1 transition the maximum value of P is 0.26.

As the incident light is absorbed essentially in the almost bulk-like intrinsic regions of the 60 nm wide GaAs layer above and below the doping spike, which is evident from PLE spectroscopy [30], excitation with, e.g., right circularly polarized light generates heavy-holes with spin -3/2 and spin -1/2 electrons as well as spin -1/2 light-holes and spin +1/2 electrons. The probabilities for these transitions are 3:1 leading to a majority of spin -1/2 electrons with an initial degree of polarization of 0.5 [37,45]. Since the number of photogenerated holes is small compared to the doping level ($\simeq$ 1 %), the spin polarization of these holes is irrelevant and therefore neglected. The radiative recombination of the spin-polarized electrons involves quantized hole states with the heavy-hole-light-hole degeneracy removed. Therefore the maximum polarization of the emitted light is also 0.5. Recombination with holes in the hh1 subband appears as a maximum in the difference spectrum I^+-I^-. Recombination with light-holes results in the emission of photons with circular polarization opposite to that from heavy-hole recombination and appears therefore as a minimum in the difference spectrum [Fig. 17(b)].

Time-resolved PL spectra provide direct information on the electron recombination and spin dynamics. The rate equations, which describe the population change of spin-up electrons (N^+) and spin-down-electrons (N^-) or, i.e., electrons with spin aligned parallel and antiparallel to the propagation direction of the exciting light, are given by [38]

$$dN^+/dt = - N^+/\tau_{rec} - N^+/\tau_{sp} + N^-/\tau_{sp} \qquad (1)$$

$$dN^-/dt = - N^-/\tau_{rec} - N^-/\tau_{sp} + N^+/\tau_{sp}. \qquad (2)$$

Here a δ-shaped excitation pulse is assumed generating the initial spin-up and spin-down electron populations $N^+(0)$ and $N^-(0)$, respectively. τ_{rec} and τ_{sp} are the recombination time and electron spin relaxation time. The change in the total

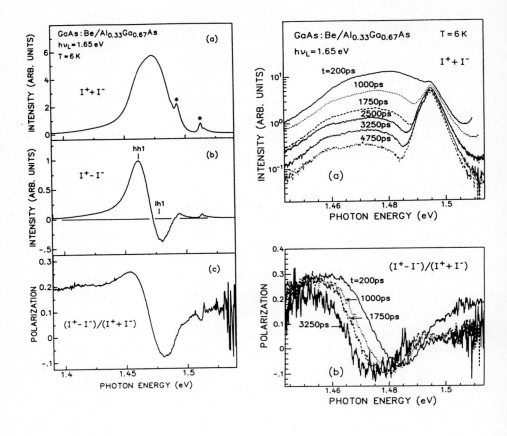

Fig. 17 (left) Low-temperature PL spectra of a δ-doped GaAs:Be/Al$_{0.33}$Ga$_{0.67}$As double-heterostructure with a dopant concentration of 8×10^{12} cm^{-2}. The spectra were recorded at 6 K with excitation at 1.65 eV. (a) Total intensity spectrum I$^+$+I$^-$, (b) difference spectrum I$^+$-I$^-$, and (c) polarization spectrum (I$^+$-I$^-$)/(I$^+$+I$^-$). Subband energies, calculated for a dopant spread of 2 nm, are marked in (b) with the position of the first heavy-hole level fixed at the peak of the hh1 emission band (Ref. 46).

Fig. 18 (right) Time-resolved PL spectra of a δ-doped GaAs:Be/Al$_{0.33}$Ga$_{0.67}$As double-heterostructure with a dopant concentration of 8×10^{12} cm^{-2} for various time delays t. The spectra were recorded at 6 K with excitation at 1.65 eV. (a) Total intensity spectra I$^+$+I$^-$ and (b) polarization spectra (I$^+$-I$^-$)/(I$^+$+I$^-$). The long-lived emission at 1.49 eV in (a) is due to donor-to-acceptor recombination in the GaAs buffer layer (Ref. 46).

population $N(t) = N^+(t) + N^-(t)$ is given by a single exponentially decaying function

$$N(t) = [N^+(0) + N^-(0)] \exp(-t/\tau_{rec}). \qquad (3)$$

The temporal evolution of the degree of spin polarization P_{sp} defined by $P_{sp} = (N^+ - N^-)/(N^+ + N^-)$ is given by

$$P_{sp}(t) = P_{sp}(0) \exp(-2t/\tau_{sp}) \qquad (4)$$

with $P_{sp}(0) = [N^+(0) - N^-(0)]/[N^+(0) + N^-(0)]$.

Thus, with the emission intensties I^+ and I^- being proportional to the electron populations N^+ and N^-, the recombination time τ_{rec} and the electron spin relaxation time τ_{sp} can be extracted from the time-resolved total intensity spectra $(I^+ + I^-)$ [Fig. 18(a)] and polarization spectra $(I^+-I^-)/(I^+ +I^-)$ [Fig. 18(b)]. At 6 K the recombination shows a biexponential behaviour with an initial time constant $\tau_{rec} = 2$ ns followed by a somewhat slower decay time of 3.5 ns. For the spin relaxation time a significantly larger value of $\tau_{sp} = 20 \pm 4$ ns is obtained. With increasing time delay the 2DHG emission shifts to lower energies. This is explained as follows. After the exciting laser pulse, photogenerated carriers reduce the spatial modulation of the band edge energies. With the concentration of those carriers decreasing with increasing time delay, the spatial modulation increases and causes a red shift of the emission from recombination of spatially separated photocreated electrons and the 2DHG.

The initial degree of polarization P_0 for the hh1 transition is 0.3. This value is lower than the maximum achievable polarization of 0.5, which indicates that some spin relaxation takes place while the photogenerated electrons relax in energy and get spatially separated from the photoexcited holes. Taking into account the reduction factor [37] $\tau_{sp}/(\tau_{rec} + \tau_{sp})$, which relates the initial polarization P_0 to the time-averaged one P_{av} observed in the cw measurement, one thus obtains an expected averaged value of $P_{av} = 0.27$ in good agreement with the measured cw polarization of $P_{av} = 0.26$ [see Fig. 17(c)].

The present spin relaxation time $\tau_{sp} = 20$ ns is two orders of magnitude longer than the relaxation time of $\simeq 200$ ps found in homogeneously p-type doped GaAs with a comparable acceptor concentration of 2.8×10^{19} cm^{-3} [43]. From work on homogeneously doped samples, it has been concluded that the BAP mechanism is relevant for electron spin relaxation in degenerate p-type GaAs at low temperatures [42,43]. For this mechanism the spin relaxation rate $1/\tau_{sp}$ in a degenerate semiconductor is proportional to $(\Delta_{exch})^2$ [42-44]. Δ_{exch} is the exchange splitting of the exciton ground state and is given by [47]

$$\Delta_{exch} = \Delta_{exch}(3D) \cdot |\Phi^{2D}(0)|^2 / |\Phi^{3D}(0)|^2 \cdot \int (X_e(z) \cdot X_h(z))^2 \, dz \quad (5)$$

where $\Delta_{exch}(3D)$ is the exchange splitting of the 3D exciton, $\Phi^{2D}(0)$ and $\Phi^{3D}(0)$ are, respectively, the 2D and 3D exciton wavefunction amplitudes at zero electron-hole relative coordinate, and $X_e(z)$ and $X_h(z)$ are the electron and hole subband envelope wavefunctions. Ignoring quantum size effects, we can write $|\Phi^{2D}(0)|^2 / |\Phi^{3D}(0)|^2 = 8a_0$, where a_0 is the 3D exciton radius (15 nm for GaAs). Based on the scheme presented in Fig. 8 one can calculate the electron-hole wavefunction overlap in equation (5). These calculations yield a two orders of magnitude decrease in $1/\tau_{sp}$, in agreement with the above described experiment. This agreement confirms that electron-hole scattering with a simultaneous exchange interaction (the BAP mechanism) is the dominant electron spin relaxation process in highly p-type doped GaAs [46].

When neglecting excitonic effects, electron spin relaxation times $\geq$ 10 ns are expected only for hole concentrations below 10^{16} cm^{-3} [43,44]. However, it has been shown that exciton formation leads to a drastic reduction of the spin relaxation time due to the enhanced electron-hole overlap [11,39]. Taking a hole concentration of about 2×10^{16} cm^{-3} as a limit where exciton formation becomes important, it follows that spin relaxation times are always $\leq$ 4 ns in homogeneously doped GaAs or in p-type doped quantum wells. In the p-type δ-doped double-heterostructures, in contrast, where exciton formation is suppressed by the built-in electric fields of the structure (Fig. 8), at least one order of magnitude longer relaxation times can be realized. This allows one to investigate electron spin kinetics in semiconductors, even in the limit of weak electron-hole interaction with the 2DHG necessary for probing the electron spin distribution.

5. Summary

We have discussed the use of photoluminescence and photoluminescence excitation spectroscopy for the study of modulation- and δ-doped semiconductor heterostructures. In these heterostructures low-dimensional electron or hole gases can be formed, which have significant effects on the interband optical properties of these structures. Many-body effects lead to an enhancement of the strength of optical transitions involving states close to the Fermi edge. In both types of heterostructures space charge induced potentials lead to a spatial separation of majority carriers and photogenerated minority carriers and thus to a reduction of the electron-hole wavefunction overlap. This wavefunction overlap can be probed optically using circular polarized light by the measurement of electron spin relaxation times in, e.g., p-type δ-doped double-heterostructures.

Acknowledgments: The author wants to thank D. Richards, H. Schneider, M. Ramsteiner, and P. Koidl for many valuable discussions and contributions to the major part of the results discussed above, A. Fischer and K. Ploog for providing high-quality sample structures, as well as H.S. Rupprecht for continuous support and encouragement.

References:

[1] See, e.g., selected chapters in "Physics and Applications of Quantum Wells and Superlattices", E.E. Mendez and K. von Klitzing (eds.), (Plenum, New York, 1987), NATO Adv. Sci. Inst. Ser. B 170 (1987).

[2] See, e.g., D. Heitmann in "Physics and Applications of Quantum Wells and Superlattices", E.E. Mendez and K. von Klitzing (eds.), (Plenum, New York, 1987), NATO Adv. Sci. Inst. Ser. B 170, 317 (1987).

[3] See, e.g., A. Pinczuk and G. Abstreiter in "Light Scattering in Solids V", M. Cardona and G. Güntherodt (eds.), (Springer, Berlin, 1989) p. 153; A. Pinczuk in "Advances in Solid State Physics", Vol. 32 (Vieweg, Braunschweig, 1992) p. 45.

[4] See, e.g., C. Weisbuch in "Physics and Applications of Quantum Wells and Superlattices", E.E. Mendez and K. von Klitzing (eds.), (Plenum, New York, 1987), NATO Adv. Sci. Inst. Ser. B 170, 261 (1987).

[5] C.H. Chang, S.A. Lyon, and C.W. Tu, Appl. Phys. Lett. 53, 258 (1988).

[6] I.V. Kukushkin, K. von Klitzing, and K. Ploog, Phys. Rev. B 37, 8509 (1988).

[7] I.V. Kukushkin, K. von Klitzing, K. Ploog, and V.B. Timofeev, Phys. Rev. B 40, 7788 (1989).

[8] S. Schmitt-Rink, C. Ell, and H. Haug, Phys. Rev. B 33, 1183 (1986).

[9] M.S. Skolnick, J.M. Rorison, K.J. Nash, D.J. Mowbray, P.R. Tapster, S.J. Bass, and A.D. Pitt, Phys. Rev. Lett. 58, 2130 (1987).

[10] G. Livescu, D.A.B. Miller, D.S. Chemla, M. Ramaswamy, T.Y. Chang, N. Sauer, A.C. Gossard, J.H. English, IEEE J. Quantum Electron. QE-24, 1677 (1988).

[11] See, e.g., T.C. Damen, L. Vina, J.E. Cunningham, J. Shah, and L.J. Sham, Phys. Rev. Lett. 67, 3432 (1991) and references therein.

[12] E. Burstein, Phys. Rev. 93, 632 (1954).

[13] Q.X. Zhao, J.P. Bergman, P.O. Holtz, B. Monemar, C. Hallin, M. Sundaram, J.L. Merz, and A.C. Gossard, Semicond. Sci. Technol. 5, 884 (1990); J.P. Bergman, Q.X. Zhao, P.O. Holtz, B. Monemar, M. Sundaram, J.L. Merz, and A.C. Gossard, Phys. Rev. B 43, 4771 (1991).

[14] A.J. Turberfield, S.R. Haynes, P.A. Wright, R.A. Ford, R.G. Clark, J.F. Ryan, J.J. Harris, and C.T. Foxon, Phys. Rev. Lett. 65, 637 (1990); B.B. Goldberg, D. Heiman, A. Pinczuk, L. Pfeiffer, and K. West, Phys. Rev. Lett. 65, 641 (1990).

[15] D. Heiman, A. Pinczuk, B.S. Dennis, L.N. Pfeiffer, and K.W. West, Phys. Rev. B 45, 1492 (1992).

[16] See, e.g., K. Ploog, M. Hauser. and A. Fischer, Appl. Phys. A 45, 233 (1988).

[17] See, e.g., F. Koch and A. Zrenner, Mater. Sci. Eng. B 1, 221 (1989).

[18] J. Wagner, M. Ramsteiner, D. Richards, G. Fasol, and K. Ploog, Appl. Phys. Lett. 58, 143 (1991); D. Richards, J. Wagner, M. Ramsteiner, U. Ekenberg, G. Fasol, and K. Ploog, Surf. Sci. 267, 61 (1992).

[19] A. Zrenner, F. Koch, and K. Ploog, Inst. Phys. Conf. Ser. 91, 171 (1988); R.B. Beall, J.B. Clegg, and J.J. Harris, Semicond. Sci. Technol. 3, 612 (1988).

[20] J. Wagner, M. Ramsteiner, W. Stolz, M. Hauser, and K. Ploog, Appl. Phys. Lett. 55, 978 (1989).

[21] N. Schwarz, F. Müller, G. Tempel, F. Koch, and G. Weimann, Semicond. Sci. Technol. 4, 571 (1989).

[22] G. Abstreiter, R. Merlin, and A. Pinczuk, IEEE J. Quantum Electron. QE-22, 1771 (1986).

[23] C.H. Perry, K.S. Lee, W. Zhou, J.M. Worlock, A. Zrenner, F. Koch, and K. Ploog, Surf. Sci. 196, 677 (1988).

[24] J.C.M. Henning, Y.A.A.R. Kessner, P.M. Koenraad, M.R. Leys, W. van der Vleuten, J.H. Wolter, and A.M. Frens, Semicond. Sci. Technol. 6, 1079 (1991).

[25] J. Wagner, A. Fischer, and K. Ploog, Phys. Rev. B 42, 7280 (1990).

[26] J. Wagner, A. Fischer, and K. Ploog, Appl. Phys. Lett. 59, 428 (1991).

[27] D. Richards, J. Wagner, A. Fischer, and K. Ploog, to appear in Appl. Phys. Lett.

[28] A.M. Gilinsky, K.S. Zhuravlev, D.I. Lubyshev, V.P. Migal, V.V. Preobrazhenskii, and B.R. Semiagin, Superlattices and Microstructures 10, 399 (1991).
[29] J. Wagner, A. Ruiz, and K. Ploog, Phys. Rev. B 43, 12134 (1991).
[30] D. Richards, J. Wagner, H. Schneider, G. Hendorfer, M. Maier, A. Fischer, and K. Ploog, to appear in Phys. Rev. B.
[31] W. Chen, M. Fritze, A.V. Nurmikko, D. Ackley, C. Colvard, and H. Lee, Phys. Rev. Lett. 64, 2434 (1990).
[32] W. Chen, M. Fritze, A.V. Nurmikko, M. Hong, and L.L. Chang, Phys. Rev. B 43, 14738 (1991).
[33] J.F. Mueller, Phys. Rev. B 42, 11189 (1990).
[34] P. Hawrylak, Phys. Rev. B 44, 6242 (1991).
[35] T. Ogawa and T. Takagahara, Phys. Rev. B 43, 14325 (1991).
[36] J.M. Calleja, A.R. Goni, B.S. Dennis, J.S Weiner, A. Pinczuk, S. Schmitt-Rink, L.N. Pfeiffer, K.W. West, J.F. Müller, and A.E. Ruckenstein, Solid State Commun. 79, 911 (1991); Surf. Sci. 263, 346 (1992).
[37] R.R. Parsons, Phys. Rev. Lett. 23, 1152 (1969).
[38] R.J. Seymour and R.R. Alfano, Appl. Phys. Lett. 37, 231 (1980).
[39] A. Tackeuchi, S. Muto, T. Inata, and T. Fuji, Appl. Phys. Lett. 56, 2213 (1990).
[40] S. Bar-Ad and I. Bar-Joseph, Phys. Rev. Lett. 68, 349 (1992).
[41] G. Fishman and G. Lampel, Phys. Rev. B 16, 820 (1977).
[42] A.G. Aronov, G.E. Pikus, and A.N. Titkov, Zh. Eksp. Teor. Fiz. 84, 1170 (1983) [Sov. Phys. JETP 57, 680 (1983)].
[43] K. Zerrouati, F. Fabre, G. Bacquet, J. Bandet, J. Frandon, G. Lampel, and D. Paget, Phys. Rev. B 37, 1334 (1988).
[44] G.L. Bir, A.G. Aronov, and G.E. Pikus, Zh. Eksp. Teor. Fiz. 69, 1382 (1975) [Sov. Phys. JETP 42, 705 (1976)].
[45] M.I. D'yakonov and V.I. Perel', Zh. Eksp. Teor. Fiz. 60, 1954 (1971) [Sov. Phys. JETP 33, 1053 (1971)].
[46] J. Wagner, H. Schneider, D. Richards, A. Fischer, and K. Ploog, to be published.
[47] See, e.g., Y. Chen, B. Gil, P. Lefebre, and H. Mathieu, Phys. Rev. B 37, 6429 (1988); U. Rössler, S. Jorda, and D. Brodio, Solid State Commun. 73, 209 (1990).

Electronic Structure and Electrical Characterisation of Semiconductor Heterostructures

N.J.Pulsford
Max-Planck-Institut für Festkörperforschung,
Heisenbergstr. 1, W-7000 Stuttgart 80, Germany

Abstract The electronic structure and electrical properties of low-dimensional semiconductor heterostructures are reviewed. Perpendicular and parallel transport in GaAs/GaAlAs heterostructures is discussed, as well as novel quantum phenomena induced by strong magnetic fields and lateral confinement.

1 Introduction

These lectures are intended to provide an introduction to the electrical properties of low-dimensional semiconductor heterostructures. In tackling such a broad field, it is not possible to cover every aspect in one short article, and I have chosen to describe some main topics to give an overview of the progress during the last decade. The first two sections cover vertical transport in tunnelling devices and superlattices, the third section discusses the use of modulation doping to achieve high mobilities in heterojunctions and the last two sections explore quantum transport in magnetic fields and point contacts. Before starting, it is assumed that the reader has a knowledge of semiconductor physics (as given for example in Ashcroft and Mermin [1]) and is familiar with the concept of confinement in heterojunctions, quantum wells and superlattices. Detailed reviews of confinement can be found in a number of texts, including the excellent books by Bastard [2] and by Weisbuch and Vinter [3]. Much of the discussion is based on the effective mass approximation which gives a physical feeling to the problems. Hopefully only a few equations appear in the text; detailed mathematical treatments of particular areas can be found in the reference list.

2 Resonant Tunnelling Devices

Tunnelling through a potential barrier is a manifestation of the quantum mechanical nature of particles and has no classical analogue. Originally applied to describe the time decay of quasi-bound states (for example, the emission of alpha particles by heavy nuclei), tunnelling found an early place in semiconductor heterostructures with the work on double barrier diodes by Chang, Esaki and Tsu in 1974 [4]. An attractive feature of tunnelling through layered semiconductors is that the tunnel barriers are similiar to those found in quantum

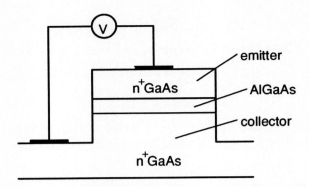

Figure 2.1: Schematic diagram of a single barrier resonant tunnelling diode.

mechanics textbooks: one-dimensional square or triangular shaped barriers typically a few penetration lengths wide, which makes a physical picture of the tunnelling process readily accessible.

Fig 2.1 shows a schematic diagram of a resonant tunnelling semiconductor diode with a single tunnel barrier. A thin undoped AlGaAs layer separates two n^+ doped GaAs contact regions (the emitter and collector). Due to the different bandgap energies in GaAs and AlGaAs, the conduction bandedge forms a potential barrier for the perpendicular motion of the conduction electrons through the device (Fig 2.2). Both the GaAs contact regions are doped and conduct so that an emitter-collector bias voltage applied to the device is dropped

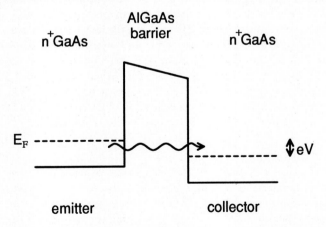

Figure 2.2: Conduction bandedge profile for a single barrier diode.

predominantly across the AlGaAs barrier. Provided that the barrier is thin enough, electrons may then tunnel through to give a current. For a typical barrier $\sim 300 meV$ high (x=0.3 Al content), the decay length $\alpha \approx 13 \text{\AA}$ which is a convenient length scale to grow barriers a few α's wideby atomic layer epitaxy (see the contribution by Klaus Ploog in these proceedings).

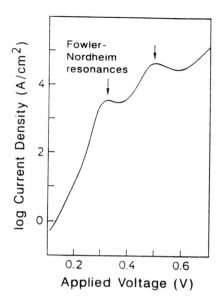

Figure 2.3: Calculated tunnelling current for 100Å (x=0.2) AlGaAs barrier [5].

Modulating the bias voltage in Fig 2.2 modulates the transmission through the barrier and produces the non-linear current-voltage characteristic which we are interested in exploiting in making a device. A simple expression for the tunnelling current through the barrier is derived as follows [5]: from the current density $j = nev_g$ where the group velocity $v_g = (1/\hbar)\partial E/\partial k$, the current from the emitter to the collector is

$$J_{EC} = \frac{e}{4\pi^3\hbar} \int dk_z dk_\parallel^2 f_E(E)[1 - f_C(E)]\frac{\partial E}{\partial k_z}T(E_z) \tag{2.1}$$

where the integration is over the density of states in k-space in the perpendicular (z) and in-plane ($\parallel$) directions. $f_i(E)$ is the Fermi-Dirac distribution function in the emitter/collector and $T(E)$ is the transmission coefficient. The in-plane density of states is

$$dk_\parallel^2 = \frac{2\pi m^*}{\hbar^2}dE_\parallel \tag{2.2}$$

The net current J equals the difference between the currents flowing in each direction

$$J = \frac{em^*}{2\pi^2\hbar^3} \int dE_z dE_\parallel [f_E(E) - f_C(E)]T(E_z) \tag{2.3}$$

which is equivalent to saying that the current is determined by the convolution of the difference in the occupation of emitter and collector states with their transmission probability. Both these terms increase with bias V giving a corresponding rise in J [5] (Fig 2.3). At high biases, the barrier is reduced to a triangle, allowing both decaying and travelling waves in the AlGaAs layer and weak resonances appear in J due to interference effects. This behaviour, often termed resonant Fowler-Nordheim tunnelling, is analogous to optical interference in thin films. Hickmott el al. reported Fowler-Nordheim resonances in the tunnelling current across single 250Å and 300Å AlGaAs barriers [6].

The Fowler-Nordheim resonances in the tunnelling current across a single barrier are normally weak because the transmission coefficient T(E) at high bias remains close to unity even when off-resonance. Much stronger resonant tunnelling features are observed in double barrier structures. The band profile of a double barrier diode is shown in Fig 2.4 and the transmission coefficient

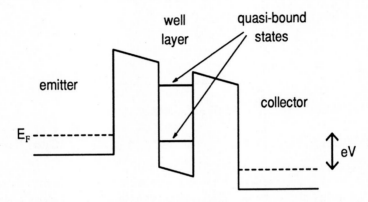

Figure 2.4: Conduction bandedge profile for a double barrier diode.

through two 50Å (x=0.3) barriers separated by a 60Å well is plotted in Fig 2.5 [5]. The sharp resonances are assigned to quasi-bound states and broader Fowler-Nordheim resonances are present at higher energy. Each quasi-confined state is characterised by a lifetime τ due to the finite probability of leaking out through the barriers, where τ is related to the resonance width ΔE by the uncertainty principle $\tau = \hbar/2\Delta E$. The longer lifetime of the lower, more strongly confined states is reflected by a sharper resonance. The tunnel current J can be calculated directly from from Eq 2.3, but it is more intuitive to consider the effect of conserving energy and in-plane momentum ($k_\parallel$) during the tunnelling process. The quantisation of k_z defines a single in-plane dispersion for each quasi-confined state and the tunnel current is derived by matching this dispersion onto the occupied emitter states (and empty collector states) with the same $k_\parallel$ and E. A schematic diagram of the resulting tunnel current is shown in Fig 2.6. Relative to the emitter, the energy of the quasi-confined state decreases with applied bias. Assuming small biases and symmetric barriers, it is given to a first approximation by $E_c = E_0 - \frac{1}{2}eV$. No resonant current can flow until E_c drops below the emitter Fermi energy; the subsequent triangular shape of the tunnel current is then a consequence of the *constant* 2D density of states as more of the $k_\parallel$ well states become available for tunnelling. The peak in J occurs when E_c aligns with the emitter band edge; below the band edge,

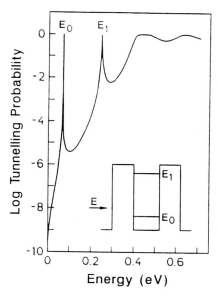

Figure 2.5: Transmission probability for two 50Å (x=0.2) AlGaAs barriers separated by a 60Å GaAs well [5].

all the $k_\parallel$ tunnelling channels are lost and J drops to zero. The abrupt drop in J gives a sharp negative differential resistance feature which is exploited in most device applications. Experimental data taken from Mendez et al. [5] are shown in Fig 2.7. Resonant peaks corresponding to each quasi-confined state are superimposed on a rising background of non-resonant tunnelling through the whole structure. Note that the triangular shape of the resonant tunnelling current is reproduced. In principle, the voltage of the resonant peaks allow a direct spectroscopic measure of the confinement energies. However care is required in modelling the peak voltages – some of the potential is dropped across the contacts (from charge depletion at the interfaces) and charge build-up in the well distorts the linear potential profile (see below).

An important parameter in device applications is the value of the peak-to-valley current ratio, describing the strength of the negative differential resistance. Peak-to-valley ratios of up to 3.9 have been reported in GaAs/AlGaAs [7] and up to 30 in the AlInAs/GaInAs structures [8] (both at room temperature). However, these values are smaller than predicted by resonant tunnelling theory. In real systems, scattering by impurities, phonons etc. perturbs the electron phase coherence and changes the tunnelling behaviour. The transit time of the electrons through the double barrier structure is limited by the lifetime τ of the quasi-confined state in the well. Clearly, if τ is comparable to the inelastic scattering time, then there is a change from a resonant to a sequential tunnelling

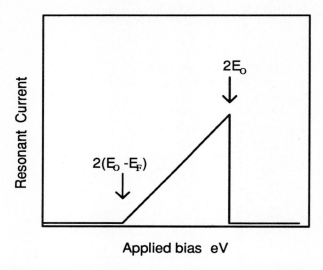

Figure 2.6: Schematic diagram of resonant current through a quasi-confined state at E_0.

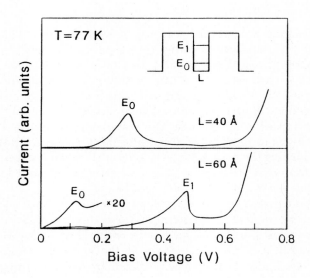

Figure 2.7: Experimental current-voltage characteristics through two double barrier diodes of different well widths. The (x=0.4) AlGaAs barriers are $100 Å$ wide in both samples [5].

(two-step) process, i.e. the phase coherence of the electrons is not maintained through the structure. However even in this limit, negative differential resistance is still observed due to the modulation of the transmission coefficient into the well. The difference arises in the size of the peak current: the transmission no longer approaches unity as for a coherent resonant process (see Fig 2.5), but is of the order of the transmission through the second barrier. From the size of the peak current, it should be therefore possible to tell whether the tunnelling is mostly resonant or sequential [5]. However in addition to reducing the peak current, scattering processes are also found to increase the non-resonant valley current. If the electrons are able to scatter, they may enter a quasi-confined state non-resonantly, with the excess momentum/energy absorbed by the impurity, phonon etc. An example of such a process is the presence of optical phonon replicas of the resonant tunnelling peak in the valley current [9]. The peak-to-valley ratios observed experimentally can be accounted for if the effects of scattering processes on the tunnelling current are considered.

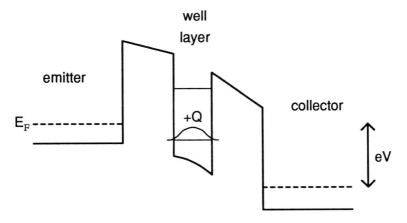

Figure 2.8: Conduction bandedge profile showing the band-bending due to the charge build up at resonance.

Resonant tunnelling devices have been developed as high frequency oscillators, with the bistability induced by linking the negative differential resistance to an external capacitive circuit [10]. However, it is also possible for a device to develop its own intrinsic bistability due to charge build-up in the well layer [11]. The intrinsic instability can be estimated as follows. Taking the lifetime of the quasi-confined state τ as the transit time through the structure, there is a charge build-up in the well layer at resonance given by $Q = J\tau$. Mendez shows that for a symmetric double barrier structure, Q is equivalent to filling the 2D states in the well up to E_F [5], and this density of charge can make a significant contribution to the potential profile. In effect, the double barrier acts like a double capacitor with a charge Q in the central plate reducing the potential drop across the first barrier and increasing the drop across the second barrier by $\frac{1}{2}Q/C$ (Fig 2.8). The reduction in the potential drop across the first barrier

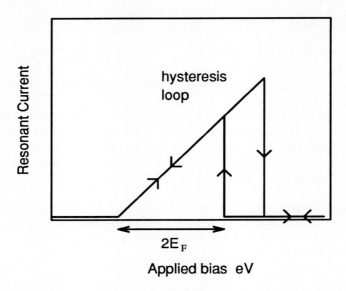

Figure 2.9: Schematic diagram showing the hysteresis in the resonant tunnelling current due to charge build-up in the well layer.

acts to maintain the tunnelling resonance with a high current density and large Q. Therefore on the up-sweep (increasing V) with Q building up gradually, a larger bias is required to reach the cut-off point, whereas on the down-sweep, the normal resonance condition applies. As illustrated in Fig 2.9, this leads to a hysteresis loop in the $J(V)$ curve. Experimentally, the hysteresis is often washed out by layer fluctuations or current oscillations associated with the external circuit. It may also be suppressed by designing an asymmetric double barrier structure with a thin collector barrier to minimise the charge build-up. If, on the other hand, the collector barrier is chosen to be very thick (to enhance the build-up) then a large hysteresis loop can be achieved [11].

3 Miniband Transport in Superlattices

A superlattice is a semiconductor structure composed of a *periodic* arrangement of thin layers of different materials (Fig 3.1). The periodicity is important because, rather than thinking in terms of the electrons tunnelling between adjacent layers, the motion through the superlattice layers is described by a coherent E-k miniband dispersion. The formation of minibands is analogous to the formation of bands in a bulk crystal – in both cases, waves can travel without dissapation through the periodic lattice so long as they do not satisfy the Bragg condition. The difference is that the length scale in a superlattice is longer, so that the minibands are formed on a smaller energy scale ($meV's$ compared to $eV's$ in a bulk crystal). In addition, for a layered structure, the minibands are only one-dimensional. Superlattices are viewed as being technologically important from the possibility to tune the miniband structure via the layer thicknesses, so-called

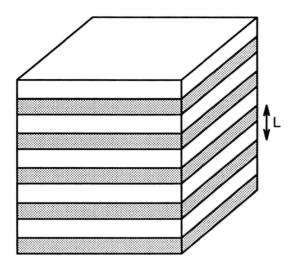

Figure 3.1: Schematic diagram showing alternate layers of two different semiconductor materials in a periodic superlattice structure.

'designer band structure', to meet specific material needs. One proposed application, based on the electrical properties of minibands is to use a superlattice as a Bloch oscillator to produce a tunable high frequency resonator [12].

The principle of a Bloch oscillator is seen from the periodic band dispersion. In Fig 3.2, the solid line shows the miniband E-k dispersion of a superlattice and the dashed line plots the group velocity of the lowest miniband derived from $v_g = (1/\hbar)\partial E/\partial k$. As the superlattice Brillouin zone boundary is approached, the gradient and hence the group velocity tend to zero, i.e. electrons suffer Bragg reflections from the superlattice layers. The semiclassical equation of motion for the electrons in an electric field E is

$$\hbar \frac{dk}{dt} = -eE \qquad (3.1)$$

which in a constant electric field gives

$$k(t) = k(0) - eEt/\hbar \qquad (3.2)$$

Therefore in the absense of scattering, the electron wavevector increases linearly with time. Oscillations arise because on passing through the Brillouin zone boundary, the electrons are reflected back $k_{BZ} \rightarrow -k_{BZ}$ (electron states separated by a reciprocal lattice vector are equivalent). Thus $v_g(k)$ oscillates with a period

$$T = \left(\frac{2\pi}{L}\right)\left(\frac{\hbar}{eE}\right) \qquad (3.3)$$

The motion in free space has the same frequency and this behaviour is known as Bloch oscillations. In any real material, scattering from impurities, phonons

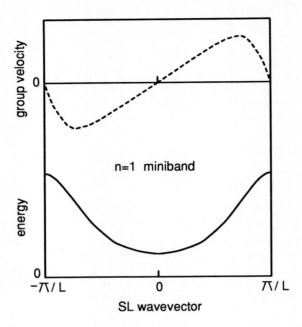

Figure 3.2: Schematic diagram of the n=1 miniband dispersion and the corresponding group velocity for a superlattice.

etc. limits the continuous increase in k and the requirement of phase coherence for the Bloch oscillations implies that the oscillatory period should be shorter than the scattering time, i.e. $T<\tau$.

We first consider the possibility of observing Bloch oscillations in bulk GaAs. For a crystal period $L = 6\text{Å}$, Eq 3.3 gives an oscillatory period $T = 10^{-11} sec$ in an electric field $E = 10 kV/cm$. A typical scattering time in undoped pure GaAs is $\tau = 10^{-13} sec$ – hence in this electric field, Bloch oscillations are not observed. In addition, the presence of multiple X,L valleys close to the Brillouin zone boundary complicates the behaviour; in very strong electric fields, intervalley scattering can dominate over Bragg reflections and suppress any oscillatory motion of the electrons [13]. A superlattice on the other hand is a more promising candidate: the miniband dispersion is simpler due to the one-dimensional potential and the longer superlattice period (up to $\sim 200\text{Å}$) reduces the Bloch oscillator period. However it is also expected that imperfections in the superlattice structure may increase the carrier scattering rate.

The influence of scattering on the current flow through a superlattice was first investigated by Esaki and Tsu using a classical approximation [12]. Their calculations showed that for a scattering rate τ^{-1}, the drift velocity v_d first increases linearly with electric field according to Eq 3.2 and then reaches a maximum at $eE\tau L/\hbar = 1$, beyond which there is a gradual decrease in v_d, i.e. a region of negative differential velocity. The decrease arises from the equilibrium population of carriers near the Brilouin zone boundary with small values of v_d.

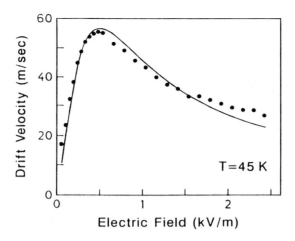

Figure 3.3: Experimental data (points) and modelled fit (line) for the drift velocity in the n=1 miniband of a 118Å GaAs / 16Å AlAs superlattice [14]. The electric field is calculated assuming a linear potential drop across the undoped region of the device.

In measurements on narrow miniband superlattices (miniband width $\sim 1meV$), Grahn et al. observe Esaki-Tsu negative differential velocity using optical excitation and time resolved transport to measure v_d [14]. Their experimental data in Fig 3.3 is fitted using an extension of the Esaki-Tsu model which differentiates between energy and momentum relaxation.

To see how this behaviour relates to the appearance of Bloch oscillations, we note that the criterion for negative differential velocity, $eE\tau L/\hbar > 1$ or $2\pi\tau/T > 1$ is less stringent than for Bloch oscillations $\tau/T > 1$. In other words, the region where Bloch oscillations should occur is at higher electric fields. In the experiments of Grahn et al. however, the miniband structure of the superlattices breaks down before the condition for Bloch oscillations is reached [14]. The strong electric field allows electrons to tunnel resonantly into the higher confined states of adjacent layers and it is no longer possible to describe the carrier transport by the coherent miniband dispersion of the lowest state. An alternative way to enhance the Bloch oscillations is to decrease the carrier scattering rate. A scattering time $\tau = 157 fsec$ is derived from the modelled fit in Fig 3.3 for the transport through the superlattice layers [14]. This scattering time is similiar to that measured in undoped bulk GaAs, indicating that scattering from superlattice imperfections is not significant in these samples. Even so, the condition for Bloch oscillations is not reached. It has been proposed to reduce the scattering rate by growing superlattices as thin vertical wires to remove the phase space for in-plane scattering events [15]. The suppression of optical phonon scattering has been observed using a strong magnetic field to quantise the in-plane dispersion of a standard superlattice [16], however no results have been reported for superlattice wire structures.

4 Modulation Doped Heterojunctions

Semiconductor transistor devices often require a high density, highly mobile channel of electrons for their operation – the high density allows a large current to be carried with little dissapation and the high mobility ensures a fast switching time. The combination of high carrier density and high mobility cannot be achieved in bulk GaAs due to the various scattering mechanisms present. For a high purity, bulk GaAs sample shown in Fig 4.1 [17], the low tempera-

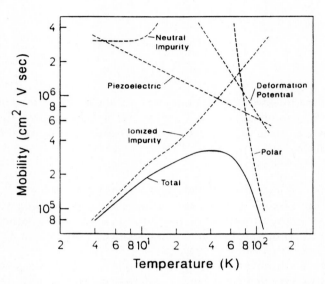

Figure 4.1: Calculated contributions to the mobility for the various scattering processes in high purity bulk GaAs [17].

ture mobility is limited by ionised impurity scattering, in the temperature range $T=50-100K$, acoustic phonon scattering via the piezoelectric and deformation potential coupling becomes important and above $T=100K$, optical phonon scattering dominates. The peak mobility is high in this sample due to the low impurity concentration; however consequently, also the carrier density is very low. The carrier density can be increased by doping, but this occurs only at the expense of the the mobility (for example with $N_D \sim 10^{17} cm^{-3}$, $\mu \sim 10^3 cm^2/V sec$ at $T=77K$) due to the large increase in scattering from ionised impurities.

The problem of impurity scattering is overcome in a modulation doped heterojunction by spatially separating the electrons from the impurity donor atoms and this allows both a high density and a high mobility to be achieved [18]. The structure of a GaAs/AlGaAs heterojunction is illustrated in Fig 4.2. Electrons from the shallow Si donors in the AlGaAs transfer across to the GaAs side to

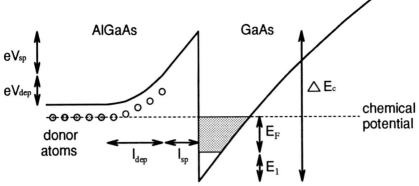

Figure 4.2: Schematic diagram of the band profile in a GaAs/AlGaAs heterojunction. The 2D electron layer is formed at the heterojunction interface.

equalise the chemical potentials in the two materials and become confined at the heterointerface in a quasi two-dimensional (2D) sheet. The confining triangular potential well for the electrons is formed by the electrostatic field of the charge transfer and the conduction band edge discontinuity. Note that at the heterointerface, the electrons are separated from the ionised dopant atoms by an undoped AlGaAs spacer layer. The charge transfer is self-consistent: as more electrons transfer across, the band bending increases, the confinement becomes stronger and the 2D electron Fermi level rises. Equilibrium is established when the chemical potentials on each side of the heterojunction are equal. In a simple model, this can be written

$$\Delta E_C = E_1 + E_F^{2D} + E_D + eV_{dep} + eV_{sp} \qquad (4.1)$$

where the various terms are defined in Fig 4.2. For a charge transfer giving a 2D electron density N_S at the heterointerface, the depletion width of the doped AlGaAs layer is $l_{dep} = N_S/N_D$ where N_D is the 3D donor concentration in the AlGaAs. From Gauss's Law, the electric field at the interface is $E_S = eN_S/\epsilon$. Solving Schrödinger's equation for an infinite triangular potential well $\phi(z) = eE_S z$ ($z > 0$) gives a confinement energy for the lowest subband

$$E_1 = \left(\frac{\hbar^2}{2m}\right)^{1/3} \left(\frac{9\pi e^2 N_S}{8\epsilon}\right)^{2/3} = \gamma N_s^{2/3} \qquad (4.2)$$

The Fermi energy of the 2D electrons is equal to N_S divided by the 2D density of states: $E_F^{2D} = \pi\hbar^2 N_S/m^*$. The potential drop in the AlGaAs layer is derived by integrating the electric field over the doped and undoped regions: $V_{dep} = \frac{1}{2}E_S l_{dep}$ and $V_{sp} = E_S l_{sp}$. Summing up these contributions gives a non-linear equation to be solved for the 2D electron density

$$\Delta E_C = \gamma N_S^{2/3} + \frac{\pi\hbar^2}{m^*} N_S + E_D + \frac{e^2}{2\epsilon N_D} N_S^2 + \frac{e^2 l_{sp}}{\epsilon} N_S \qquad (4.3)$$

In a more detailed analysis, there are additional contributions to Eq 4.3. The occupation of states at finite temperature is described by the Fermi-Dirac distribution (unimportant if $T \ll E_D, E_F^{2D}$). The depletion charge due to residual doping in the nominally undoped GaAs and AlGaAs regions adds a weak electric field to the potential profile. Electron–electron interactions in the 2D layer are included as a Hatree term $V_{ee}(z)$ in the calculation of the confinement energy E_1. However these terms normally add only small corrections to ΔE_C.

There are two important parameters which describe the 2D electrons in a heterojunction: the 2D electron density N_S and the electron mobility μ. Experimentally, electron densities in the range $N_S = 0.3$–$20 \times 10^{11} cm^{-2}$ and electron mobilities up to $\mu \sim 10^7 cm^2/V sec$ can be achieved in GaAs/AlGaAs heterojunctions. A typical electron density $N_S = 5 \times 10^{11} cm^{-2}$ corresponds to a Fermi energy $E_F^{2D} = 18 meV$, which compares with the thermal energy $kT = 0.3 meV$ at liquid helium temperatures. Hence the 2D electrons form an attractive system to work with: at low temperatures, they behave like a metal ($kT \ll E_F$, large μ) whose Fermi energy can be varied over a wide range. The electron transport can often be modelled by adapting the theories developed for bulk metals (the electron effective mass, the dielectric constant and the Fermi energy all need to be scaled). Boltzmann transport, exchange interactions, screening, etc. all become applicable. There are also a few changes introduced by the quasi 2D nature of the electrons: crystal momentum conservation is relaxed in the z-direction due to the confining potential and screening is weaker as the cloud of 2D electrons is unable to completely surround an external (3D) charge. A point which is often overlooked is the role played by the electron-electron interactions in the 2D layer. For an electron density $N_S = 5 \times 10^{11} cm^{-2}$, the average electron separation $d = 1/\sqrt{\pi N_S}$ is $d = 80\mathring{A}$ which corresponds to a Coulomb energy $e^2/4\pi\epsilon d = 14 meV$. Despite the large Coulomb energy ($\sim E_F$), the electrons can still be treated as an ensemble of *independent* single particles, because the influence of the electron-electron interactions is strongly limited by Pauli exclusion principle (a similiar effect also occurs in metals). In the single particle picture, the electron-electron interactions are approximated by a small renormalisation of the dispersion (change in energy, effective mass etc.). The name often given to the 2D electron layer: two-dimensional-electron-gas (2DEG) reflects the single particle nature of their behaviour.

It is instructive to consider the influence of the heterojunction design on the electron density and mobility. The 2D electron density (Eq 4.3) increases with the donor doping concentration and with the conduction bandedge discontinuity, though decreases with the width of the undoped AlGaAs spacer layer, as more of the bandedge discontinuity is taken up by the potential drop eV_{sp} for a given charge transfer N_S. The conduction bandedge discontinuity scales with the aluminium fraction (x) in the AlGaAs layer and can be used to increase N_S. However care is required in treating the Si donor level as the direct–indirect bandgap crossover in the AlGaAs ($x=0.37$) is approached. Shubert and Ploog [19] show that the energy of the donor level (E_D) increases from $6meV$ for $x<0.2$ to more than $100meV$ for $x>0.3$. At the same time, an increasing proportion of the donors become attached to a deep level some $\sim 150meV$ below the AlGaAs bandedge. The rapid increase in E_D and the presence of deep level impurities are detrimental to achieving a high 2D electron density and can even lead to carrier freeze-out at low temperatures. The effect of the segregation of dopant atoms during growth has also been identified [20]. On a conventionally grown heterojunction (first GaAs, then AlGaAs), this segregation produces a spatially dependent donor concentration, with a higher density of donors located away from the heterointerface. As the 2D electron density is determined by donors near the heterointerface, this can lead to a lower electron density than expected from the nominal doping concentration. Segregation may be suppressed by reducing the growth temperature or by choosing a lower aluminium content.

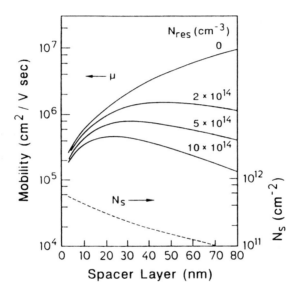

Figure 4.3: Calculation of the mobility and electron concentration as a function of spacer layer width in an AlGaAs heterojunction for various residual impurity concentrations in the GaAs buffer layer [21].

A high electron mobility can be maintained in a heterojunction for a wide range of 2D electron concentrations due to the spatial separation of the donor atoms from the 2D electron channel. However the precise mobility is determined by the interplay of various factors. Hirakawa and Sakaki [21] consider the scattering in GaAs/AlGaAs heterojunctions by acoustic phonons, optical phonons, the low density of residual ionised impurities in the GaAs and the high density of ionised donor impurities introduced into the AlGaAs. Phonon scattering is due to bulk GaAs phonons as there is little wavefunction penetration into the AlGaAs layer and confined interface modes are not important. The ionised impurity scattering is described by the Born approximation which assumes that it is both elastic and weak. In this model, the scattering cross-section for an ionised impurity located a distance z' from the heterointerface is

$$\sigma(\theta) = \int dz |\psi(z)|^2 exp[-2k|z' - z|sin(\theta/2)] \tag{4.4}$$

where $\psi(z)$ is the confined electron wavefunction and k is the electron wavevector ($\sim k_F$ as only electrons near the Fermi surface have the available phase space to scatter). Screening by the 2D electrons is included by a wavevector dependent dielectric constant. Fig 4.3 shows the calculated mobility as a function of spacer layer thickness for different residual GaAs impurity concentrations. The AlGaAs donor concentration is fixed at N_D=3.4x10^{17}cm^{-3} and the temperature is low so that phonon scattering is weak. The mobility peaks at an intermediate spacer layer thickness which contrasts with the 2D electron concentration which shows a monotonic decrease (dashed line). For a thin spacer layer, the mobility is limited by scattering from the high density of AlGaAs donor atoms and scales as the inverse of N_D. For wider spacer layers, the scattering is weaker from the AlGaAs donor atoms (larger z' in Eq 4.4) and the mobility increases, until scattering from the residual impuritites in the GaAs becomes important. z' in Eq 4.4 is now fixed and the dependence comes from $k \sim k_F$, which decreases with the electron density. Correspondingly, the mobility decreases with the spacer layer width and is inversely proportional to N_{res} in this regime. By modelling a set of samples with different spacer layer thicknesses, Hirakawa and Sakaki are able to determine N_D and N_{res} from the dependence of the mobility [21]. Note that for a given doping concentration, the maximum mobility is limited by the concentration of residual impurities and this is normally the factor which limits the mobility in very high quality samples.

The influence of phonon scattering is seen in the temperature dependence of the mobility. Data for a series of heterojunctions are shown in Fig 4.4 [22]. At high temperatures, the curves are similiar to those measured for bulk GaAs (Fig 4.1). Above $T = 100K$, optical phonon scattering dominates and in the temperature range $T = 50-100K$, acoustical phonon scattering through the piezoelectric and deformation potential coupling contributes. The difference lies in the low temperature mobility, where there is only a weak dependence for the 2DEG in a heterojunction compared to the strong decrease measured in bulk GaAs. However this just reflects the change in system from an undoped semiconductor ($\mu \sim T^{\frac{3}{2}}$) to a metal ($\mu \sim T^0$) for the ionised impurity scattering. In the case of phonon scattering, the strong temperature dependence of the phonon population makes this difference less obvious. As the heterojunction quality and hence the electron mobility increases, there is also a systematic change in the low temperature gradient of the mobility from being weakly positive to weakly negative [23]. The positive gradient occurs for strong ionised impurity scattering and is a second order effect arising from the energy dependence of the scattering rate (through k in Eq 4.4). This adds a small positive quadratic term to the mobility. The negative gradient is observed in higher quality samples where the ionised impurity scattering is weaker. Then, even though the acoustical phonon scattering still only makes a small contribution to the overall mobility, its temperature dependence can be felt and adds a small negative linear term to

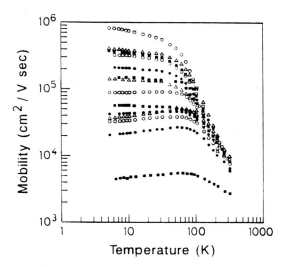

Figure 4.4: The mobility measured as a function of temperature in a series of GaAs/AlGaAs heterojunction samples [22].

the mobility. Finally in discussing the mobility of heterojunctions, two more contributions should be noted. Firstly, interface roughness increases the scattering rate, though this is normally only important for samples grown outside the optimum temperature range. Secondly, the occupation of a higher subband increases the scattering due to the extra phase space available. The latter effect can be studied systematically by using a top gate [24] or parallel magnetic field [25] to depopulate higher subbands.

5 The Shubnikov-de Haas and Quantum Hall Effects

The application of a magnetic field has proved a powerful tool for the study of the Fermi surfaces of metals, those of copper and aluminium being good examples. In the previous section, the 2DEG in a heterojunction was shown to be metallic at low temperatures with a well-defined Fermi surface ($kT \ll E_F \sim 1\text{--}50 meV$). This section concentrates on the use of a magnetic field to probe the 2D electrons. Starting from the classical theory for electron conduction in a magnetic field, the effect of Landau quantisation of the 2D electron energy states is developed. The analogy with metals is pushed to the limit with the cyclotron energy becoming as large as, or even greater than the 2D Ferm energy.

The explanation of the Hall effect is a success of the classical theory of conduction in solids. By matching the Lorentz force on a current flow with the electric field due to charge build-up, the Hall voltage across a sample is

$$U_H = \frac{B_z I_x}{l N_{3d} e} \tag{5.1}$$

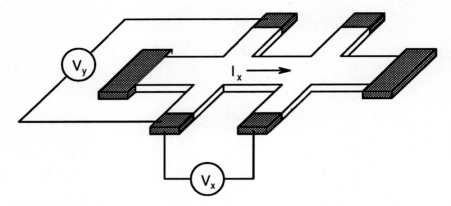

Figure 5.1: Schematic diagram of a Hall bar to measure V_x and V_y.

where B_z is the magnetic field, I_x is the current, l is the sample thickness and N_{3d} is the (3D) electron density. This formula is extended to a 2D system by noting that lN_{3d} is equivalent to a 2D charge density N_S projected onto a plane perpendicular to the magnetic field direction. Defining the Hall resistance as $R_H = U_H/I_x$ gives

$$R_H = \frac{B}{N_S e} \qquad (5.2)$$

It is remarked that Eq 5.2 does not depend on any of the physical dimensions of the sample. For a 2D system, the electric field is related to the current density by a 2x2 resistivity tensor

$$\underline{E} = \underline{\rho} \cdot \underline{j} \qquad (5.3)$$

Defining the current flow in the x-direction only, this reduces to

$$E_x = \rho_{xx} j_x \qquad (5.4)$$

$$E_y = \rho_{xy} j_y \qquad (5.5)$$

with the classical forms for the resistivity tensors being

$$\rho_{xx}(B) = \frac{1}{N_S e \mu} \qquad (5.6)$$

$$\rho_{xy}(B) = \frac{B}{N_S e} \qquad (5.7)$$

where the expression for ρ_{xx} comes from the definition of the mobility. Eqns 5.6 and 5.7 are only valid assuming $j_y = 0$, which experimentally requires a Hall bar sample with separate current and voltage probes (Fig 5.1). A quantitative comparison is made between the classical values (dashed lines) and the measured values (solid lines) for ρ_{xx} and ρ_{xy} in Fig 5.2. As the magnetic field is increased, clear deviations from the classical behaviour are observed. The longitudinal resistivity ρ_{xx} oscillates with a period ($\sim N_S/B$) and the Hall resistance ρ_{xy}

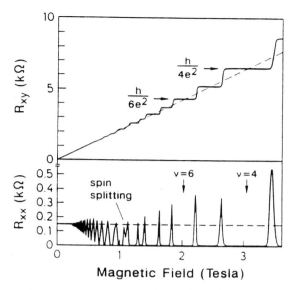

Figure 5.2: Comparison between the classical values (dashed lines) and measured values (solid lines) for the longitudinal and Hall resistance of a GaAs/AlGaAs heterojunction at $T=50mK$. The 2D electron concentration is $3\times 10^{11} cm^{-2}$ and the mobility is $2.8\times 10^5 cm^2/Vsec$ [26].

has well-defined plateaus which align with the minima in ρ_{xx} (the quantum Hall effect [27]). Remarkably, the quantised values of the Hall resistance plateaus do not depend on the sample parameters (or even material) and are given by h/ie^2 where i is an integer to an accuracy better than 1 in 10^7. This high degree of accuracy has lead to the quantum Hall resistance being adopted as the resistance standard with $R_K = h/e^2$ defined as 25812.807Ω so that the Ω retains (as near as possible) its original value [28].

Anyone familiar with Fermi surface studies of metals will recognise the periodic $(1/B)$ oscillations in the resistivity tensors as due to the Landau quantisation of the magnetic orbits, which modulates the density of states at the Fermi energy. Landau quantisation is valid for $\omega_c\tau > 1$ which even for modest mobilities $\mu = 10^5 cm^2/Vsec$, is achieved below $B = 1T$. For a 2D system, the Hamiltonian in a magnetic field is

$$H = \frac{1}{2m^*}(\underline{p} - e\underline{A})^2 + V(z) \qquad (5.8)$$

where $\underline{A}$ is the magnetic potential and $V(z)$ is the confinement potential. If the magnetic field is in the z-direction, the Hamiltonian separates into in-plane motion in the magnetic field and perpendicular confinement. The eigenvalues in this case are

$$E = E_z + (n + \frac{1}{2})\hbar\omega_c \qquad (5.9)$$

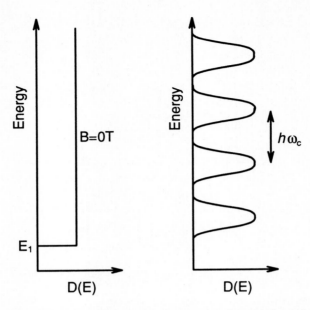

Figure 5.3: Schematic diagram of the change in the density of 2D states into Landau levels in a magnetic field. Only the lowest subband is shown.

E_z is the subband energy and $\hbar\omega_c$ is the cyclotron energy. In an ideal system, the energy states are a series of equally spaced delta functions, however these are broadened in a real system by scattering from phonons, impurities etc. The resultant density of states is illustrated in Fig 5.3. The degeneracy of each Landau level is $eB/\hbar$ per unit area (which as a rule of thumb, is equal to the 2D density of states over an energy interval $\hbar\omega_c$). It is convenient to characterise the system by the number of occupied Landau levels for a given magnetic field B

$$\nu = \frac{hN_S}{eB} \qquad (5.10)$$

where ν is termed the filling factor. With increasing magnetic field, the filling factor decreases as the highest Landau level successively depopulates due to the increase in degeneracy of the lower levels. Thus, the Fermi energy oscillates from being where the density of states is high in the centre of a Landau level to where the density of states is low between two levels and the conductivity (proportional to the square of the density of states) oscillates accordingly. Including the twofold spin degeneracy, an integer number of Landau levels is occupied for $\nu = 2, 4, 6, 8...$ (note also that integer values of ν leads to the 1/B periodicity). In relating the conductivity to the resistivity, it is important to follow the tensor relation between $\underline{j}$ and $\underline{E}$ (Eq 5.3)

$$\rho_{xx} = \frac{\sigma_{xx}}{\sigma_{xx}^2 + \sigma_{xy}^2} \qquad (5.11)$$

If σ_{xx} is small, i.e. $\sigma_{xx} \ll \sigma_{xy}$, then $\rho_{xx} \sim \sigma_{xx}/\sigma_{xy}^2$. This result that the resitivity is proportional to the conductivity is due to the influence of the strong Hall field and the way that the conductivity and resistivity tensors are defined.

The simple model explains oscillations in ρ_{xx} arising from the modulation of the density of states at E_F, with minima occuring at even integer filling factors when an integer number of Landau levels are fully occupied. However the experimental results go further (see Fig 5.2). In strong magnetic fields, ρ_{xx} vanishes for wide magnetic field ranges and the spin splitting of individual levels is clearly resolved. The vanishing conductivity implies no states at the Fermi energy for wide field ranges. The density of states between Landau levels can be very low, but due to the continuous depopulation, the Fermi energy jumps between Landau levels as the magnetic field is increased. The simple model cannot account for the broad minima in ρ_{xx}. In addition, it cannot explain the plateaus in ρ_{xy}. Calculating ρ_{xy} for an integer filling factor $\nu = i$ gives

$$N_S = \frac{ieB}{h} \quad (5.12)$$

$$\rho_{xy} = \frac{B}{N_S e} = \frac{h}{ie^2} \quad (5.13)$$

which are the observed values of the Hall plateaus. However these values are derived using magnetic fields at which an integer number of of Landau levels are fully occupied. That is, Eq 5.13 just defines a series of points which lie on the straightline classical value of the Hall resistance (Eq 5.7). We are interested in what happens away from these field values. In a classical picture, there is no change in N_S with B and hence there is no reason to think that ρ_{xy} should deviate from its linear increase and give well-defined plateaus.

The plateaus in ρ_{xy} are observed in samples with a wide range of mobilities, however surprisingly, they are best resolved in the samples with an intermediate mobility $\mu \sim 10^5 cm^2/Vsec$. In higher quality samples, there is a general trend of plateau width decreasing with mobility. This dependence suggests that impurity scattering leading to Landau level broadening is important. Indeed numerical calculations show that the impurity broadening introduces a difference between the states at the centre of the level and the states in the tails (at the edge): only those in the centre are extended and can carry current, whereas those in the tails are strongly localised [29]. A convenient picture is to think of the attractive and repulsive interactions of the impurities as producing a random electrostatic potential 'landscape' at the heterojunction interface. Semiclassically, the electrons move on magnetic orbits around equipotential contours of the undulating landscape. Orbits in the Landau level tails are at high or at low energies, and therefore correspond to closed orbits around potential 'hills' or in potential 'lakes' – i.e. they are strongly localised. Only those orbits in the centre of the level are sufficiently spatially extended to contribute to the current flow.

The vanishing ρ_{xx} can now be explained as occuring when the Fermi energy lies in the Landau level tails where the states are localised. The lack of conducting states at the Fermi energy gives the $\sigma_{xx}=0$ but the finite density of localised states allows the conductance minimum to extend over a finite magnetic field range (the ratio of localised to extended states, i.e. the sample quality, can be estimated from the relative width of the ρ_{xx} minima). It is also clear from Fig 5.2 that the ρ_{xy} plateaus are the same width as the ρ_{xx} minima and hence also reflect the presence of localised states. In a semiclassical picture, only extended states in the centre of a Landau level carry current and contribute to the Hall voltage. Depopulating localised states in a Landau level tail increases the number of occupied extended states in lower Landau levels (degeneracy $\propto$ B); the increase in 'current carrying' N_S cancels the increase in B in Eq 5.7 and leads to a constant Hall resistance. However, although explaining the existence of Hall plateaus, this still does not account for the quantisation at precise values of h/ie^2 because not all the electrons are in current carrying states. Semiclassically there is no simple explanation for the precise quantisation. Quantum mechanical calculations show that the quantisation is due to the extended states compensating for the localised states in the current flow [30]. However this is not very satisfying as no symmetry can be invoked, although an analogy can be made with the increased rate of flow of water around an obstacle in a pipe. A more complete explanation can be found in the edge-state model which treats the current flow as skipping magnetic orbits, i.e. 1D channels, at the edge of the sample. The Hall plateaus then arise from the quantisation of the 1D conductance (see next section). However even this picture breaks down in the high current regime when the Hall voltage becomes greater than the cyclotron energy. The reader is referred to recent reviews for a complete description of the edge-state model [31].

The spin slitting of the Landau levels is clearly resolved in Fig 5.2, even though the bare Zeeman energy (g^*=0.33) is $\sim$ 100 smaller than the cyclotron energy (m^*=0.07). Analysis of the oscillations in ρ_{xx} by rotating the sample so that the Zeeman energy ($\propto B$) coincides with the cyclotron energy ($\propto$ the perpendicular component of B) reveals an enhancement of the g-factor up to g^*=6 at odd integer filling factors [32]. This enhancement is explained by the Hatree-Foch exchange interaction which has the following mechanism [33]. Due to the Pauli exclusion principle, electrons of like-spins cannot occupy the same region of space and hence by 'avoiding' each other, they reduce the strength of their mutual Coulomb interaction. Electrons of opposite spins experience the full Coulomb interaction. Hence for unequal spin populations (odd ν), there is a reduction in Coulomb energy of the lower spin state which appears as an increase in the spin splitting. When both spin states are equally occupied (even ν), the Coulomb interactions are the same for both spin states and the spin splitting is not 'enhanced'. The enhancement of the spin splitting is thus a quantum mechanical manifestation of the strong Coulomb interactions between the electrons in the 2D layer, which were mentioned in the previous section.

As the sample quality is increased, new ρ_{xx} minima and ρ_{xy} plateaus appear between the integer filling factors at lower temperatures. Magnetotransport traces are shown in Fig 5.4 for a sample with a mobility μ=2.1x10$^6 cm^2/V sec$ measured at T=30mK [34]. Remarkably, the new minima/plateaus appear exactly at odd fractional values of the filling factor $\nu = \frac{p}{q}$ where p,q are integers with q odd and hence are referred to as the fractional quantum Hall effect [35].

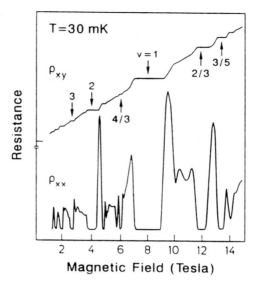

Figure 5.4: Magnetotransport traces for a low density, high mobility sample in the regime of the fractional quantum Hall effect [34]

In contrast to the 'integer' quantum Hall effect, the fractional features become stronger as the sample quality increases indicating that they are an intrinsic property of the 2D electrons and not related to localisation effects. Present theories relate the fractional states to electron-electron interactions in the magnetic field which condense the electrons into a stable many-body ground state when a Landau level is fractionally filled [36]. The presence of only odd denominator fractions is accounted for by the anti-symmetry requirement of the ground state wavefunction and the ρ_{xx} minima and ρ_{xy} plateaus appear because the ground state is separated from the excited states by a finite energy gap, similiar to the gap between extended states in the integer quantum Hall effect. However even though the effect on ρ_{xx} and ρ_{xy} is similiar, the underlying reason for the energy gap is clearly different in the two cases.

6 Quantum Point Contacts

The 2D nature of the electrons in a heterojunction arises from the quantisation of the perpendicular motion into discrete electric subbands by the interface potential. The motion in the plane is decoupled and remains essentially 'free'. However further quantisation into 1D or 0D subbands is possible by applying a suitable lateral potential to the heterojunction plane. The first such structure in which 1D conduction effects were resolved was the quantum point contact which is described in this section [37, 38]. An advantage of working with a 2D heterojunction is that planar semiconductor technology can be used to achieve the lateral definition (see the contribution by S.Radelaar in these proceedings).

In any structure, the relative length scales determine whether quantum confinement effects can be observed. For a 2D electron layer in a GaAs heterojunction with $N_S = 3 \times 10^{11} cm^{-2}$ and a mobility $\mu = 2 \times 10^6 cm^2/V\,sec$, the electrons at the Fermi surface have a wavelength $\lambda_F = 45 nm$ and move with a velocity $v_F = 2.3 \times 10^5 m/sec$. Their mean free path between collisions is $\lambda \approx 20 \mu m$. The mean free path has important implications for transport in small structures. For example, in a wire with a lateral width $L \sim \lambda$, scattering at the wire boundaries becomes important and the concept of a mean free path between diffusive scattering events is no longer valid [39]. In this regime of quasi-ballistic transport, the electron motion can often be described by classical 'billard ball' dynamics. However, in order to observe lateral quantum confinement effects, which we are interested in here, it is necessary to go to even smaller length scales, such that the width of the potential modulation is the order of λ_F ($\ll \lambda$).

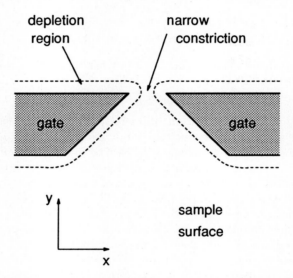

Figure 6.1: Schematic diagram of the metal gate and depletion region to form a quantum poit contact.

Conduction through 1D subbands was first resolved in quantum point contacts formed by evaporating a metal gate on to the surface of a heterojunction [37, 38]. The structure of the gate in the region of the point contact is shown in Fig 6.1. The width of the constriction through the metal gate is $\sim 250 nm$ with a length $\sim 50 nm$. Applying a negative bias voltage to the gate depletes the region immediately under the metal with only the narrow constriction left free; increasing the gate voltage decreases the size of the constriction until eventually pinch-off is achieved leaving the two sides of the 2DEG electrically insulated. In this way, the width of the constriction may be continuously tuned by the gate voltage. The conductance between the emitter and collector as a function of the gate voltage (constriction width) is shown in Fig 6.2 [37]. Rather than dropping smoothly to zero as would be expected classically, the conductance shows a series of well-defined steps at quantised values of

$$G = \frac{2e^2}{\hbar}i \tag{6.1}$$

where i is an integer. The data in Fig 6.2 is measured at $T=300mK$; increasing the temperature to a few Kelvin, the plateaus acquire a finite slope and gradually lose their resolution. We first present a simple theory to explain the origin of these steps and then discuss the influence of the gate geometry.

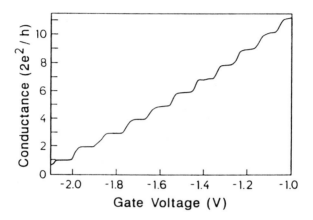

Figure 6.2: The measured conductance as a function of gate voltage for a quantum point contact at $T= 300mK$ [37].

The conductance steps are interpreted in terms of the depopulation of the 1D subbands formed by the narrow constriction [37, 38]. The constriction acts like a bottleneck for the electron motion and therefore determines the conductivity. The conductance is modelled by treating the two wide 2DEG regions (the emitter and collector) either side as two contact reservoirs, each in thermal equilibrium at chemical potentials μ_1 and μ_2 connected by 1D subbands in the narrow constriction. Electrons passing through the constriction are assumed to be in thermal equilibrium with the reservoir from which they originate and hence the 1D subbands in one direction are populated up to μ_1 and in the other direction up to μ_2 (see Fig 6.3). The difference in populations leads to a net current given by

$$I = \frac{e}{2} \sum_{n=1}^{N} \int_{\mu_1}^{\mu_2} dE \frac{dN_n}{dE} v_n(E) T_n(E) \tag{6.2}$$

where the sum is over all the occupied 1D subbands. The factor $\frac{1}{2}$ arises because the density of states (dN_n/dE) is resolved into k-motion in opposite directions.

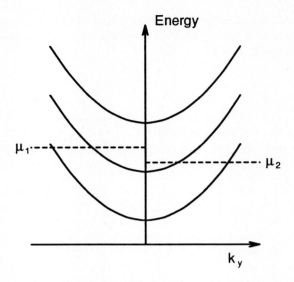

Figure 6.3: Diagram of the 1D subband dispersions in a point contact under bias. The difference in chemical potentials leads to a net current flow.

$v_n(E)$ and $T_n(E)$ are the group velocity and the transmission for the nth subbands respectively. Note that Eq 6.2 is similiar to the formula for the tunnelling current through a single barrier (Eq 2.1). The density of states for a 1D subband is

$$\frac{dN_n}{dE} = \frac{2}{\pi}\left(\frac{dE_n}{dk_y}\right)^{-1} \quad (6.3)$$

(there are $2/\pi$ k_y states per unit length). The group velocity is defined as

$$v_n = \frac{1}{\hbar}\frac{dE_n}{dk_y} \quad (6.4)$$

which leads to the result that the product of the group velocity and the density of states

$$v_n \frac{dN_n}{dE} = \frac{4}{\hbar} \quad (6.5)$$

is constant *independent* of the energy. This is a universal result for any 1D subband; it does not depend on the form of the dispersion and the cancellation is exact. The same result does not hold for 2D or 3D subbands because all the states with a finite in-plane wavevector also contribute to the conductance, 'washing out' the steps. For small bias voltages, $\mu_2 - \mu_1 = eV \ll E_F$, then

$$G = \frac{2e^2}{h}\sum_{n=1}^{N} T_n(E_F) \quad (6.6)$$

If it assumed that no backscattering takes place (see below), then $T_n(E_F) = 1$ and the conductivity is quantised

$$G = \frac{2e^2}{h} N \qquad (6.7)$$

depending on N, the number of 1D subbands occupied in the narrow constriction. As the bias voltage is increased, the 1D subbands depopulate and quantised steps appear in the conductivity. The steps are of equal height because the injected current is equipartitioned amongst the 1D subbands, with each subband carrying an equal amount of current (this is the consequence of the cancellation of the group velocity and the density of states for motion in 1D).

It is immediately clear that the quantised units of the 1D conductance are the same as the units of the quantised Hall effect and hence these conductance steps have been hailed as a quantum Hall effect without magnetic field. However the resolution of the conductance plateaus is much poorer, and can only be measured with an accuracy of $\sim 1\%$ (which compares with an accuracy of 1 in 10^7 for the quantum Hall plateaus). One problem is that to obtain the 'correct' quantised step values experimentally, a series resistance must be subtracted from the measured conductance. The series resistance arises from the resistance of the wide 2DEG regions and from the Maxwell spreading resistance which expresses the change from a narrow constriction to a wide 2D electron layer [37]. However this is typically only a small correction and independent of the applied gate voltage. The sharpness and flatness of the plateaus is also found to depend on the precise electrostatic potential in the narrow constriction [40]. Often devices of 'identical' design have different plateaus which is interpreted as due to local inhomogeneties in the doping concentration and/or the Fermi surface pinning potential in the heterojunction plane.

Although the simple 1D subband model accounts for quantised conductance steps, the success of the model is surprising considering the geometry of the point contact device [40]. Firstly, it is not expected that the 1D subbands should be well-developed in a constriction whose length L is comparable to its width W, i.e. the transmission of evanescent waves might be important. Secondly, the coupling to the wide 2DEG regions is not treated explicitly in the model (only through a series resistance); both the diffraction and the quantum mechanical reflection of the electrons are ignored. Thirdly, the emitter and collector 2DEG's are only approximations to two reservoirs in thermal equilibrium at μ_1 and μ_2; for example, scattering by impurities near the point contact may allow electrons to return through the constriction without first reaching thermal equilibrium with the intermediate reservoir.

The coupling between the wide 2DEG regions and the point contact is explained by the adiabatic nature of the gate potential. Conductance plateaus are best resolved in devices in which the constriction width changes gradually on the scale of λ_F. A 'slow' change in the constriction width implies *adiabaticity* : the confined quantum states evolve smoothly with the potential rather than suddenly changing into new states [41]. An adiabatic change involves no subband mixing and hence no reflection, only transmission through the point contact i.e.

$T_n(E)=1$. Note that the complete transmission of the 1D subbands is assumed in the derivation of Eq 6.7. If the constriction is made abrupt (for example, a narrow slit between the gates), then sharp resonance appear in the conductance steps at low temperatures. The abrupt change in the potential at each end mixes the subbands and hence partially reflects the 1D travelling waves. Standing 1D waves can build up along the constriction length and the resultant interference fringes (i.e. a modulation in the transmission) appear as sharp resonances in the conductance.

Fine structure in the conductance steps can also arise from backscattering events from impurities in and near the constriction which modulate the transmission coefficient. This is normally a small effect due to the long mean free path in the 2DEG, but may be enhanced by increasing the length of the constriction. In long constrictions ($L \sim \mu m$), conductance steps are barely resolved due to the increased backscattering by impurities and, perhaps more importantly, by channel wall irregularities [42]. Backscattering is suppressed if the constriction is made very short, but then the steps can acquire a finite slope due to transmission through evanescent waves (though the steps survive even at 'zero' length). These considerations lead to an optimum size for the channel geometry. The lateral potential should be 'smooth' to ensure adiabicity, with a lateral width and length $L \sim W \sim 100 nm$ to achieve the best step definition.

7 Acknowledgements

The author would like to thank a number of people for helpful discussions, particularly Hugues Pothier, Daniela Pfannkuche and Dieter Weiss on the role of edge-states. The author is indebted to the Royal Society for the provision of a European Science Exchange Fellowship during his stay in Stuttgart.

References

[1] N.W.Ashcroft and N.D.Mermin, "Solid State Physics", Holt-Sanders, Tokyo, Japan (1981)
[2] G.Bastard, "Wave Mechanics Applied to Semiconductor Heterostructures", Les Ulis, France (1988)
[3] C.Weisbuch and B.Vinter, "Quantum Semiconductor Structures", Academic Press, London, England (1991)
[4] L.L.Chang, L.Esaki and R.Tsu, Appl. Phys. Lett. 24, 593 (1974)
[5] E.E.Mendez in "Physics and Applications of Quantum Wells and Superlattices", NATO ASI Series B170, 159 (1987)
[6] T.W.Hickmott, P.Solomon, R.Fisher and H.Morkoc, Appl. Phys. Lett. 44, 90 (1984)
[7] M.J.Paulus, C.A.Bozada, C.I.Huang, S.C.Dudley, K.R.Evans, C.E.Stutz, R.L.Jones and M.E.Cheney, Appl. Phys. Lett. 53, 204 (1988)
[8] T.P.E.Broekaert, W.Lee and C.G.Fonstad, Appl. Phys. Lett. 53, 1545 (1988)
[9] V.J.Goldman, D.C.Tsui and J.E.Cunningham, Phys. Rev. B36, 7635 (1987)

[10] T.C.L.G.Sollner, E.R.Brown, C.D.Parker and W.D.Goodhue in "Electronic Properties of Multilayers and Low-Dimensional Semiconductor Structures", NATO ASI B231, 283 (1990)
[11] M.L.Leadbeater, E.S.Alves, L.Eaves, M.Henini, O.H.Hughes, F.W.Sheard and G.A.Toombs, Semicond. Sci. Technol. 3, 1060 (1988)
[12] L.Esaki and R.Tsu, IBM J.Dev. 14, 61 (1970)
[13] R.O.Grondin, W.Perod, J.Ho, D.K.Ferry and G.J.Iafrate, Superlattices Microstruct. 1, 183 (1985)
[14] H.T.Grahn, K.von Klitzing, K.Ploog and G.H.Döhler, Phys. Rev. B43, 12094 (1991)
[15] H.Sakaki, Jpn. J. Appl. Phys. 28, L361 (1989)
[16] H.Noguchi, T.Takamasu, N.Miura and H.Sakaki, Surf. Sci. 267, 562 (1992)
[17] G.E.Stillman and C.M.Wolfe, Thin Solid Films 31, 69 (1976)
[18] R.Dingle, H.L.Störmer, A.C.Gossard and W.Weigmann, Appl. Phys. Lett. 33, 665 (1978)
[19] E.F.Shubert and K.Ploog, Phys. Rev. B30, 7021 (1984)
[20] M.Heiblum, E.E.Mendez and F.Stern, Appl. Phys. Lett. 44, 1064 (1984)
[21] K.Hirakawa and H.Sakaki, Phys. Rev. B33, 8291 (1986)
[22] B.J.F.Lin, Ph.D.Thesis, Princeton University, Princeton, New Jersey (1984)
[23] H.W.Liu, R.Ferreira, G.Bastard, C.Delalande, J.F.Palmier and B.Etienne, Appl. Phys. Lett. 54, 2082 (1984)
[24] H.L.Störmer, A.C.Gossard and W.Weigmann, Solid State Commun. 41, 707 (1982)
[25] T.Englert, J.C.Maan, D.C.Tsui, and A.C.Gossard, Solid State Commun. 45, 989 (1983)
[26] R.J.Haug, private communication
[27] K.von Klitzing, G.Dorda and M.Pepper, Phys. Rev. Lett. 45, 494 (1980)
[28] T.Quinn, Metrolgia 26, 69 (1989)
[29] H.Aoki and T.Ando, Solid State Commun. 38, 1079 (1981)
[30] R.E.Prange, Phys. Rev. B23, 4802 (1981)
[31] M.Büttiker in "Semiconductors and Semimetals" 35, 191 (1992)
[32] R.J.Nicholas, R.J.Haug, K.von Klitzing and G.Weimann, Phys. Rev. B37, 1294 (1988)
[33] T.Ando and Y.Uemura, J. Phys. Soc. Jpn. 37, 1044 (1984)
[34] R.J.Nicholas in "Physics and Applications of Quantum Wells and Superlattices", NATO ASI Series B170, 249 (1987)
[35] D.C.Tsui, H.L.Störmer and A.C.Gossard, Phys. Rev. Lett. 48, 1559 (1982)
[36] for review see "The Quantum Hall Effect", eds R.E.Prange and S.M.Girvin, Springer-Verlag, New York (1987)
[37] B.J.van Wees, H.van Houten, C.W.J.Beenakker, J.G.Williamson, L.P.Kouvenhouven, D.van der Marel and C.T.Foxon, Phys. Rev. Lett. 60, 848 (1988)
[38] D.A.Wharam, T.J.Thornton, R.Newbury, M.Pepper, H.Ahmed, J.E.F.Frost, D.G.Hasko, D.C.Peacock, D.A.Ritchie and G.A.C.Jones, J. Phys. C21, L209 (1988)
[39] C.W.J.Beenakker and H.van Houten, Phys. Rev. B38, 3232 (1988)

[40] for review see H.van Houten, C.W.J.Beenakker and B.J.van Wees in "Semiconductors and Semimetals" 35, 9 (1992)
[41] L.I.Glazman, G.B.Lesovik, D.E.Khmel'netskii and R.I.Shekhter, JETP Lett. 48, 238 (1988)
[42] H.van Houten, C.W.J.Beenakker, P.H.M.van Loosdrecht, T.J.Thornton, H.Ahmed, M.Pepper, C.T.Foxon and J.J.Harris, Phys. Rev. B37, 8534 (1988)

New Optoelectronic Devices

Emmanuel Rosencher
Central Research Laboratory THOMSON-CSF
Domaine de Corbeville, F-91404 ORSAY Cedex

1. Introduction

The purpose of this lecture is to introduce students to the latest developments of optoelectronic devices. These devices are based on the strong quantum effects which take place in semiconductor quantum wells (QWs). The optical properties of low-dimensional structures have been adressed by J. Wagner and this topic will be mostly focused on effects due to an applied electric field on the quantum wells. Instead of providing a zoological-type classification of all the different new realizations in this field, we will mainly make a comprehensive analysis of two generic types of devices : Self Electro-optical Effect Device (SEED) and Inter-Sub-Band Transitions (ISBT) devices. The lecture is organized in the following way.

In a first part, we will briefly recall the physics of optical processes in quantum wells stressing the important concept of dipolar moment. We will introduce the difference between interband and intersubband optical transitions. In the second part, we will describe the Stark effect in interband transitions and show an important application of this phenomenon in Self Electrooptical Effect Devices. In the third part, we will adress the physics of intersubband transitions, with application related to infrared detection and modulation devices.

2. Interband and Intersubband transitions

Electrons in a semiconductor quantum are particularly well described by an effective mass Hamiltonian[1]

$$H = -\frac{\hbar^2}{2m^*}\left[\frac{\delta^2}{\delta x^2} + \frac{\delta^2}{\delta y^2} + \frac{\delta^2}{\delta z^2}\right] + V(z) \qquad (1)$$

where z represents the growth direction, h is Planck's constant and $V(z)$ is the profile of the confining potential. m^* is the effective mass of the electron in its band: We will be concerned with the conduction and the valence effective masses, m_c and m_v respectively. Without any perturbation (light, electric field, ...), the eigenfunctions $\Psi_{n,k}(r)$ and the eigenenergies e_n of the electron in the conduction band are solutions of the Schrödinger equation $H\Psi_{n,k}(r) = e_{n,k}\Psi_{n,k}(r)$ and given by :

$$\Psi_{n,k}(r) = \zeta_n(z)\, u_c(r)\, e^{i k_{//} r_{//}} \qquad (2)$$

and

$$e_{n,k} = E_n + \frac{\hbar^2}{2m_e}|k_{//}|^2 \qquad (3)$$

Here, $k_{//}$ and $r_{//}$ are the wavevector and the coordinate in the xy plane and $u_c(r)$ is the periodic part of the Bloch function in the conduction band at k=0. ζ_n and E_n are respectively the envelope wavefunction and the transverse energy of the nth subband, solutions of the one-dimensionnal Schrödinger equation $H_0 \zeta_n(z) = E_n \zeta_n(z)$ where H_0 is the z part of the Hamiltonian H in Eq. (1), i.e. $H_0 = -\hbar^2/2m_e\, d^2/dz^2 + V(z)$. The subband energy dispersion curves are given in Eq.(3) and subbands in the conduction band are schematically shown in Fig.1. The same type of equations are obained for electrons in the valence band with m_v instead of m_c.

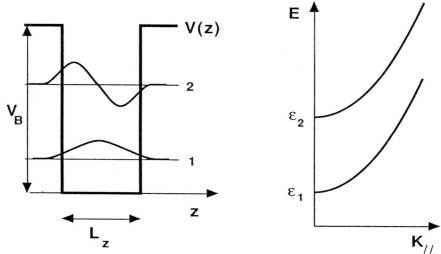

Figure 1 : A Quantum Well structure showing two subbands with their wavefunctions perpendicular to the interface and their corresponding energy spectrum.

What happens if an electromagnetic field is incident on the structure? A time dependent perturbation Hamiltonian is added in Eq.1, which is given by the dipolar interaction term $q\mathbf{E_0}\,\mathbf{r}\cos\omega t$, where q is the electronic charge. One can consider optical transitions between pairs of states |1> and |2>. The electrons in the 2-level system are polarized by the electromagnetic field, with the volume polarization given by $P = \varepsilon_0\,\chi^{(1)}\,E_0$. The quantum mechanical calculation of complex first order optical susceptibilities for a two-level system is straightforward and yields[2] :

$$\chi^{(1)} = \frac{N}{\varepsilon_0 \hbar}|\langle 1|q\mathbf{r}|2\rangle|^2 \frac{1}{(\omega_{12} - \omega) + i\Gamma/2} \qquad (4)$$

Here, N is the volumic concentration of absorbing states, $\omega_{12} = (E_2-E_1)/\hbar$ is the Bohr frequency between both states and Γ is the phase relaxation time, term equivalent to a damping constant. The susceptibility of the whole structure is thus obtained by integrating the elementary susceptibility given by Eq. 4 over all the states |1> and |2> in the band structure of the QW. The final susceptibility is related by Maxwell's equations to the macroscopic absorption coefficient α through the relations $\alpha = \omega\,\chi''\,/\,n_b c$, where c is the velocity of light, n_b is the bulk refractive index and χ'' is $-\operatorname{Im}(\chi^{(1)})$.

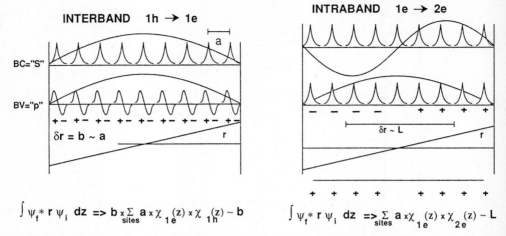

Figure 2 : Schematic of the calculation of interband (a) and intersubband (b) dipole matrix elements. In the first case, the periodic part of the Bloch functions dominate, leading to dipole with the same order of magnitude than the lattice constant. In the second case, the envelope part of the wavefuction dominates, leading to giant dipoles[3].

Two different types of behaviour regarding optical properties are obtained, depending on whether the optical transitions take place between or inside bands.

2.1. Interband transitions

In that case, the two levels belong to two different bands. The complex dipole moment μ_{12} is given by :

$$\mu_{12} = \langle 1 | qr | 2 \rangle = \int_{crystal} \psi_{n,v}^*(k,r) \, qr \, \psi_{m,c}(k',r) \, d^3r \qquad (5)$$

where v and c refer to the valence and the conduction band respectively, and n and m refer to the quantum number of the state in the respective band (n and m are equal to one if the first subbands are at stake, for instance). This dipolar moment is calculated by inserting in Eq. 5 the expressions of the wavefunctions given by Eq. 2 . Since the scale on which u_c and u_v oscillate is small (see Fig.2), this dipolar element can be written in the following form[3] :

$$\langle \Psi_{n,v}| q\mathbf{r} | \Psi_{m,c}\rangle =$$

$$\sum_i \zeta_{n,v}(z_i)\zeta_{m,c}(z_i) e^{i(\mathbf{k}_{//}-\mathbf{k'}_{//})\mathbf{r}_{i//}} \int_{cell} u_v^*(\mathbf{r'}) q(\mathbf{r'}+\mathbf{r}_i) u_c(\mathbf{r'}) d^3\mathbf{r'} \quad (6)$$

where i designates the position of each unit cell in the crystal structure and r' is the position in each unit cell. Since the Bloch parts of the wavefunctions are orthonormal, the expression above factorizes under the following form :

$$\langle \Psi_{n,v}| q\mathbf{r} | \Psi_{m,c}\rangle =$$

$$\int_{cell} u_v^*(\mathbf{r'}) q\mathbf{r'} u_c(\mathbf{r'}) d^3\mathbf{r'} \cdot \int_{QW} \zeta_n(z) \zeta_m(z) dz \cdot \delta(\mathbf{k}_{//},\mathbf{k'}_{//}) \quad (7)$$

The first term $\mu = \langle u_v | q\mathbf{r} | u_c \rangle$ represents the dipole matrix element associated to transitions within a unit cell, basically the same as in the bulk material. This matrix element has the same order of magnitude as the extension of a unit cell, i.e. ≈ 0.5 nm. The second one is the overlap integral in the quantum well (QW) between both envelope wavefunctions. It is of the order of unity for allowed transitions and zero for transitions between different parity states. The third one represents the wavevector selection rule for the optical transitions, i.e conservation of the parallel wavevector. The schematic diagram of interband transitions is shown in Fig.3a. It is clear from this figure that the complete optical susceptibility is obtained by summing up the transitions on all the states in the conduction and valence bands. Complete calculations may be obtained in usual textbooks[4]. One finally finds for the transition between the first subbands :

$$\chi(\omega) = \quad (8)$$

$$\frac{\mu^2}{\varepsilon_0 \hbar} \int_{E_g/\hbar}^{\infty} \frac{\rho_{QW}(\omega_0)}{L_z} [f_v(\omega_0) - f_c(\omega_0)] \frac{1}{(\omega_0 - \omega) + i\Gamma/2} d\omega_0$$

where ρ_{QW} is the density of states in quantum wells (for the first subband $\rho_{QW} = m_c/\pi \hbar^2$ that is 2.8×10^{13} eV^{-1}.cm^{-2}), f_c and f_v are the Fermi occupancy factor in the bands and L_z is the thickness of the QW.

The main features are shown in Fig. 3a. The absorption start at a photon energy hν given by hν = $E_g + E_{1,c} + E_{1,v}$ where E_g is the semiconductor energy gap, $E_{1,c}$ and $E_{1,v}$ are the confinement energies in the conduction and valence band respectively. It exhibits plateaus at each increase of QW density of states. The absorption photon energy is in the 1.5 - 2 eV range for GaAs, i.e. the near infrared.

2.2 Intersubband transitions.

In that case, the initial and final states are in the same band. The band indices (v=c) are then the same in Eq. 6. The dipole elements $<u_c | qr | u_c>$ are now zero by symmetry consideration (see Fig. 2) so that the total dipolar matrix elements is now given by :

$$\langle \Psi_{n,c} | qr | \Psi_{m,c} \rangle = \int_{QW} \zeta_{n,c}(z) \, z \, e_z \, \zeta_{m,c}(z) \, dz \, \delta(k_{//}, k'_{//}) \quad (9)$$

e_z is the unit vector normal to the quantum wells. Two major differences may be observed in the intersubband and interband optical transitions, as they appear in Eq. 7 and 9. First, the dipole in intersubband transitions has the same order of magnitude as the quantum well, *i.e.* few nanometers. Consequently, huge dipole moments are expected in such ISBT transitions. Secondly, the E vs $k_{//}$ initial and final dispersion curves being parallel, the transitions energy are the same for all $k_{//}$.

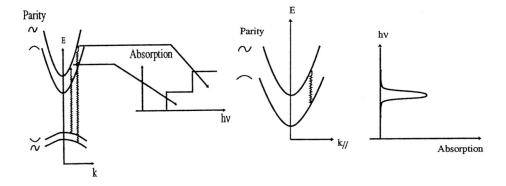

Figure 3 : (a) Schematic of interband optical transitions. The absorption spectrum is spread in energy with plateaus corresponding to the increasing of density of states in the QWs. (b) Schematics of intersubband transitions. Since the initial and final subbands are parallel, the absorption curve is sharply peaked at resonance.

The schematic diagram of intersubband transition is represented in Fig. 3b. Consequently, the absorption peaks at the transition energy given by the energy difference between the two subbands. Finally, only the component of the electric field which is normal to the layers is coupled to the QW. Normally incident light beam, for instance, cannot be absorbed. Such a selection rule does not apply in material with effective mass anisotropy such as SiGe heterostructure[5] or in hole subbands where large subband mixing between heavy and light holes occur when $k \neq 0$[6].

2.3 Excitonic effect

In fact, in interband transitions, the absorption curve described in Fig. 3b is not obtained in QW. Indeed, as is well known, a photocreated conduction electron interacts via the Coulomb interaction with the hole left behind in the valence band to yield a hydrogen-like complex, the exciton[3].

Besides providing an energy shift of the absorption equal to the exciton binding energy, the electron-hole correlation concentrates the dipole moment in the same way as quantum wells do : The electron is localized by the Coulombic potential of

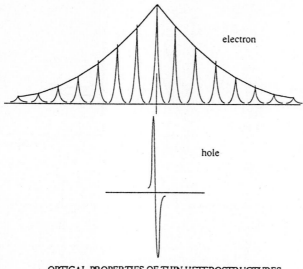

Figure 4 : (a) Electron and hole wavefunction overlap in the exciton.

OPTICAL PROPERTIES OF THIN HETEROSTRUCTURES

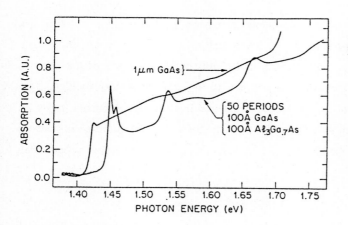

Figure 4 : (b) Comparison of the room temperature absorption spectra of bulk and 10 nm-thick QW GaAs samples. The sharp exciton peak in the MQWs denotes an enhancement of dipolar moment[7].

the hole instead of the heterostructure band offset. This is readily seen by examining the electron-hole wavefunction overlap in Fig. 4a, which is much more important than in the delocalized electron-hole pairs. The dipolar moment per crystal unit surface can be shown to be[7] :

$$\mu_{exc} = |\langle u_c | qr | u_v \rangle| \frac{8}{\pi a_B^2} \qquad (10)$$

Here, a_B is the Bohr radius in GaAs, i. e., $a_B = 4\pi\epsilon h^2 / m_{red} q^2$ where m_{red} is the reduced mass of electron and hole ($1/m_{red} = 1/m_c + 1/m_v$) and ϵ is the GaAs relative permittivity. This leads to a value for a_B of ≈ 10 nm in GaAs. Absorption exhibits a huge resonance when excitons are formed, as is shown in Fig. 4b. Let us note that the binding energy of the 2D exciton is 4 times higher than the 3D one, so that 2D excitons are observed at room temperature.

3. Stark Effect and Self Electrooptic Effect Devices[7]

If an electric field is applied to the quantum wells, the shape of the confining potential changes as well as the electron eigenfunctions and energies : this is the Stark effect [8] (see Fig. 5a). We give a rough description of this effect. The electron envelope wavefunction in the QW may be well approximated by a linear combination of the first two subband wavefunctions that we note now on |1> and |2>. One may thus apply the perturbation theory. In a symmetric quantum well, the term $< i | z | i >$ are zero and a second order development is necessary, which yields :

$$E_1(F) = E_1 - \frac{|\langle 1 | z | 2 \rangle|^2}{E_2 - E_1} q^2 F^2 \qquad (11)$$

The same expression is obtained for the holes in the valence band, so that the interband transition energy is quadratically reduced with the electric field. For electric fields in the 10^5 V/cm, quantum well thickness L_z in the 10 nm range (i.e. $z_{12} = |<1 | qz | 2>| = 16/9\pi^2 L_z$ for an infinite well ,that is, in the few nm range), this yields red shift in the few 10 meV range. Such a Stark shift in Hydrogen atoms where the dipole moment is rather in the 10^{-2} nm range, would necessitate fields as high as $10^{10} - 10^{11}$ V/cm .

This red shift, shown in Fig. 5b, can be used to make modulators for optical communication. However, the residual absorption in those devices are still rather high, being close to a resonance, and NbLiO$_3$ modulators are still prefered in most practical applications. The main application is for logical circuits which are explained below.

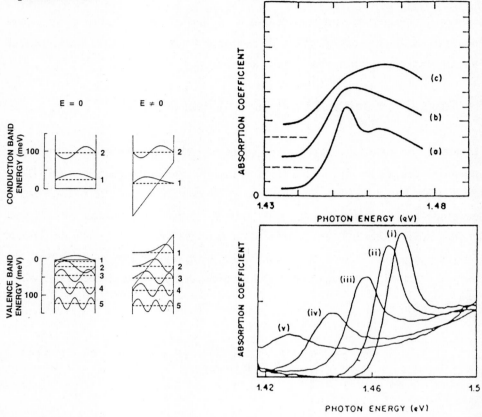

Figure 5 : (a) Schematics of the action of an electric field on a quantum well: the quadratic Stark effect. (b) Absorption spectra of a 9.4 nm QW with an applied electric field : fields are 0, 6 x 10^4, 1.1 x 10^5, 1.5 x 10^5 and 2 x 10^5 V/cm respectively[7].

One must note that such a red shift in the absorption spectra with the electric field was also observed in bulk semiconductors : It is the Franz-Keldysh effect. However, the Stark effect in QW (also named Quantum Confined Stark Effect) yields energy shifts which are much larger than in the 3D case, due to the the fact that energy levels can be strongly modified by applying large electric fields while still retaining a good oscillator strength as the electron-hole overlap remains finite because of the confinement of the electrons and holes against either interface of their common QW. An additional bonus of this confined situation is the remnant of some excitonic correlation effect in the oscillator strength even at the largest electric field whereas in 3D excitons are ionized at very moderate electric fields.

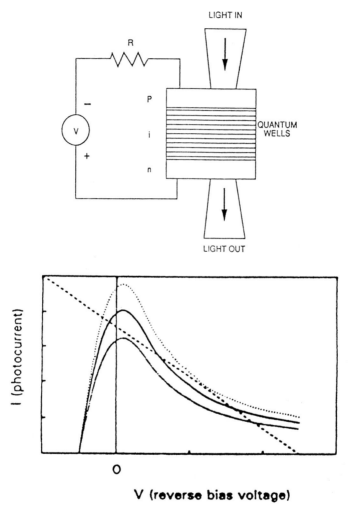

Figure 6 : (a) Schematics of a SEED biaising. (b) I-V curves in SEED device for different light power[10].

Practically, these electric fields are applied by inserting these QWs in a p-n junction (see Fig. 6a). Once the electrons have been photoexcited on the first level of the QW, they can tunnel out through the triangular potential barrier. This give rise to a photocurrent flowing from the contacts, which exhibits the same red shift with the electric field. The variation of the photocurrent with the applied bias for a given photon energy (see Fig. 6b) is the result of the competition between two effects :

-Enhancement due to the increasing quantum transparency of the triangular barrier and to the electric field drift in the barrier conduction ban,

- Decrease due to the detuning from resonance because of the red shift.

Under certain conditions, it exhibits a negative differential resistance, which is known to yield bistability if properly charged on a resistor : This is a resistor SEED. When no light is shining, all the voltage is applied to the structure, resulting in a small absorption. As light flux starts to increase, a voltage drop occurs across the resistor due to the photocurrent, reducing the voltage applied to the QWs therefore increasing absorption. Above a "switch-down" light flux, this process will lead to a runaway which puts the diode in a high absorption, high current, low applied voltage mode. This sequence of effects may be summarized as below :

$$P_{in} \to P_{abs} \to N \to I_{ph} \to P_e \to \Delta F \to \Delta E \to \Delta \alpha$$

where P_{in} is the incoming optical power, P_{abs} is the absorbed power, N the photocreated current density, I_{ph} the photocurrent, P_e the energy supplied by the electrical generator, ΔF the change in the electric field, ΔE the relative change in the energy positon of the QW subbands and α the resulting change in absorption coefficient. A sketch of the P_{out} vs. P_{in} curve (a so-called "transfer function" in logical circuits[9]) is shown in Fig. 7, together with an experimental curve.

The commutation speed of such a device is given by the product RC, where C is the capacitance and R is the dynamical resistance of the device. Operation in the 20 GHz range has been reported in such devices. The optical switching energy is ≈ (1/2) CV^2. For a typical voltage drive of 5V, the switching energy is 140 fJ, which compares well with the energy per bit (the "figure of merit") in electronic logical circuits[9].

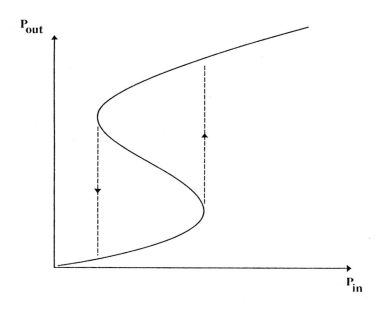

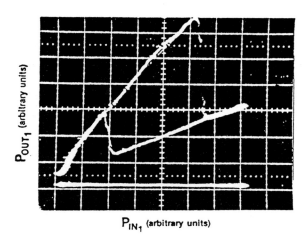

Figure 7 : Schematic of transfer function of a SEED device (a) and experimental result (b)[10].

Many implementations of SEED devices have been demonstrated. One of the drawbacks of the simple SEED is the lack of intrinsic gain (the "fan-out"[9]), necessary for cascading the bits, which requires the device to be operated near the threshold. This near threshold operation makes the device very sensitive to any fabrication fluctuations, spurious light illumination... To overcome this difficulty, symmetric SEED (S-SEED) have been developed. Integration of 2000 devices has thus been realized with 40 pW holding power per device and 1 ns of switching time. These circuits are envisioned to make parallel optical data processing in a near future[10].

4. Intersubband transition devices

The energy difference between the first two subbands in GaAs/AlGaAs is easily calculated using the Schrödinger equation and is shown in Fig. 8 as a function of QW thickness for different Al concentrations in the barrier. Because of the Γ-X cross-over above an Al concentration of 40%, the corresponding photon wavelength

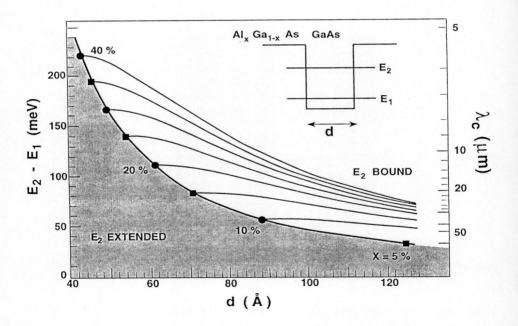

Figure 8 : Energy difference $E_2 - E_1$ between the first two bound states in a QW as a function of QW thickness for different Al composition in the barrier. The curves are bounded at their left by a continuous curve indicating the occurence of bound to extended transitions.

cannot be smaller than ≈ 5µm, so that intersubband transitions in GaAs/AlGaAs seem well adapted to optical processes in the mid-infrared. Above an Al concentration of 40%, there is a difference between the optical threshold (between Γ states) and the thermal threshold (between Γ and X states) which is detrimental to the device functioning. This limit can be elevated by using other materials (InGaAs/InAlAs on InP for instance[6]). Two different regimes are expected : Transitions between two bound states ("bound-to-bound"), which has been observed in absorption for the first time by West and Eglash[11], and transitions between a bound state and extended ones ("bound-to-extended" or photoemission). The first type of transition leads to a Lorentzian lineshape absorption, given by Eq. 4, with N replaced by the electron density ρ_s. Figure 9a shows a typical absorption curve : The measured electron phase relaxation time $1/\Gamma$ (also denotted T_2) is in the 0.1-0.5 ps range so that the typical absorption at resonance is in the 0.5% range per well.

The second type of transition leads to an absorption curve which is spread in energy, as may be seen in Fig. 9b. The absorption curves may be theoretically calculated by convolving the Lorentzian lineshape with the density of final states in the AlGaAs conduction band. Let us note that, in order to obtain absorption, the quantum well must be doped so that the electrons fill the lowest subband. Typical doping concentration is in the 10^{12} cm^{-2} range. The first application of intersubband transitions has been the fabrication of quantum detector by Levine et al.[12] The structure consists in a multiquantum well layer clad between two heavily doped GaAs ohmic contacts (see Fig. 10). The principle is the following. At sufficiently low temperature, most of the electrons are located in the confined state in the QW.

The structure is thus insulating. Upon illumination, the electrons are excited from the bound state towards the AlGaAs conduction band, where they are free to move. These photoexcited electrons are swept by the applied electric field, which yields a photocurrent. Typical response is in the 1A/W range. Figure 9b shows the comparison between the absorption and the photocurrent spectra in the same sample. The difference between both curves is still unexplained[13].

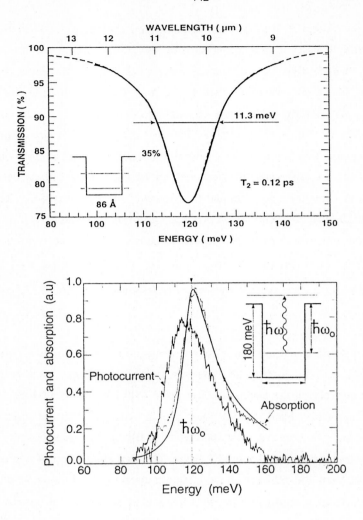

Figure 9 : (a) Experimental absorption spectrum (solid line) and theoretical fit (dashed line) in a bound-to-bound transition. (b) Experimental absorption spectra (light line) and theoretical fit (bold line) in a bound to extended transition. The photocurrent spectrum is also indicated[13].

What are the advantages of QW detectors compared to usual interband transition detectors using small gap semiconducors such that HgCdTe? Firstly, GaAs is a mature, thermally stable, highly pure material with a technology controlled in an

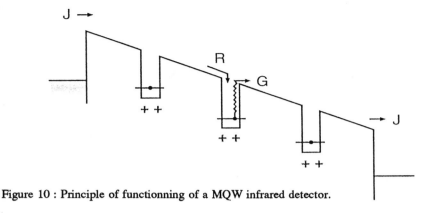

Figure 10 : Principle of functionning of a MQW infrared detector.

industrial way. Epitaxial growth allows samples with excellent uniformity over large area to be obtained, which is not the case for HgCdTe. Thus, though researches in that field has begun since only few years, large 128x128 arrays have already been realized with large yield and excellent uniformity [6]. Secondly, different wavelength may be detected on the same device, allowing multicolor detector to be fabricated. One just has to change the QW width or the Al concentration during growth. Such operations are extremely difficult to realize in other systems.

The main drawbacks are the following. Firstly, only the component of the electromagnetic field normal to the QW may be detected. Grating coupling must thus be performed, which necessitates an additional technological step. The second point is more detrimental : The working temperature of QW detectors ("BLIP" conditions for Background Limited Infrared Performances") is smaller than interband ones by typically 10 to 20K. BLIP temperatures for 10.6 μm detectors are ≈ 65 K for QW detectors and ≈ 80 K for HgCdTe. The origin of this difference lies in the different lifetime of the photoexcited electrons in their conduction band, as explained below[14].

The density of photoexcited carriers n_{3d} is given by the balance equation between recombination R and generation G_{op} rates at each quantum well (QW) . R is given by $n_{3d} / L_B \tau$, where n_{2d} is the two-dimensional density of photoexcited cariers, L_B the barrier thickness and τ is the electron lifetime in the AlGaAs barriers before recombination in the QW. G_{op} is $\alpha_{1qw} \Phi$, where α_{1qw} is the absorption efficiency of a single QW and Φ is the photon flux. Finally, the photocurrent J_{op} is given by $J_{op} = n_{3d} \, q \, v = q \, \alpha_{1qw} \Phi \, \tau \, v / L_B$, where v is the carrier velocity which may be obtained through a drift transport model in AlGaAs, e.g. $v = \mu \, F$ where μ is the mobility of carriers in the AlGaAs barrier and F is the electric field. In this model, the electrode contact is supposed to be ohmic (or injecting), that is, it supplies as many carriers as the photocurrent requires. A similar expression may be obtained for an interband photoconductor, with α_{3D} instead of α_{1QW} / L_B. The device performance is given by the signal to noise ratio, the noise being due to the dark current J_{dark} fluctuations which is controlled by the same lifetime as the photocurrent. The noise having a Poisonnian character, the noise is proportional to $\sqrt{J_{dark}}$, which yieds a $1/\sqrt{\tau}$ dependence of the signal to noise ratio. The lifetime τ is dominated by the Auger recombination time in HgCdTe, which is ≈ 1 μs, while τ is in the few ps range in QWs. The origin of such a low value is still unclear, but QW detectors must be operated at a lower temperature than HgCdTe ones, so that the dark current is reduced by the carrier freeze-out. Similar arguments show also that, though the responsivity of QW detectors is independent of the number N_{QW} of QWs, the signal to noise ratio is proportional to $\sqrt{N_{QW}}$.

One of the main advantages of QW detectors - besides the technological ones and the multispectrality potential - lies in the possibility to realize sophisticated functions by using the wealth of quantum effects in QWs. Let us go back to the Stark effect. It can be greatly enhanced in intersubband transitions by working in asymmetrical quantum wells. Figure 11 shows the band diagram of a $GaAs/Al_xGa_{1-x}As/Al_yGa_{1-y}As$ heterostructure. When an electric field is applied to the structure, the energy levels E_i are shifted by a quantity given by the first order perturbation theory, i.e. $q <i|qz|i>$. The energy difference between the first two subbands is thus :

$$\Delta E = q \, F \, \delta_{12} = q \, F (\langle 2|z|2 \rangle - \langle 1|z|1 \rangle) \qquad (12)$$

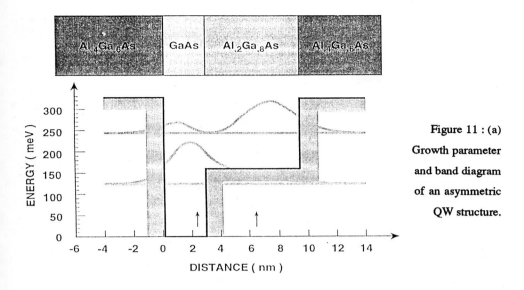

Figure 11 : (a) Growth parameter and band diagram of an asymmetric QW structure.

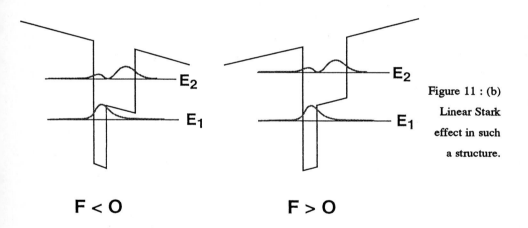

Figure 11 : (b) Linear Stark effect in such a structure.

In the structure shown in the inset of Figure 12, the value of δ_{12} is calculated to be 4.0 nm. Huge *linear* Stark shift are thus expected and measured. Figure 12 shows the normalized photoresponse of an asymmetrical MQW detector : The photocurrent peak is shifted from 8 µm up to 12 µm (that is over the whole atmospherical transparency window) by applying an electric field from -40 to +40 kV/cm. Such a device will allow the realization of solid state infrared spectrophotometers.

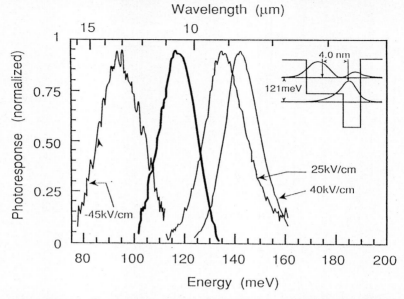

Figure 12 : Normalized responsivity of asymmetric MQW detector under positive and negative applied bias. The responsivity peak is shifted on the whole athmospheric transparancy window[17].

The same type of asymmetrical QWs may be used for optical switching in the mid-infrared, as shown by Wang and collaborators[5]. As a matter of fact, in a more general way, huge optical nonlinearities are expected in such AMQW structure. Let us recall that non-linear optical phenomena are well described by expanding the electric polarization P in powers of applied electric field E as[4] :

$$P(t) = \varepsilon_0 \chi^{(1)} \tilde{E} e^{i\omega t} + \varepsilon_0 \chi^{(2)}_{2\omega} \tilde{E}^2 e^{2i\omega t} + cc + \varepsilon_0 \chi^{(2)}_0 \tilde{E}^2 \quad (13)$$

where $\chi^{(1)}$, $\chi^{(2)}_{2\omega}$ and $\chi^{(2)}_0$ are the linear, second harmonic generation and optical rectification coefficients respectively. It may be shown that, for a doubly resonant

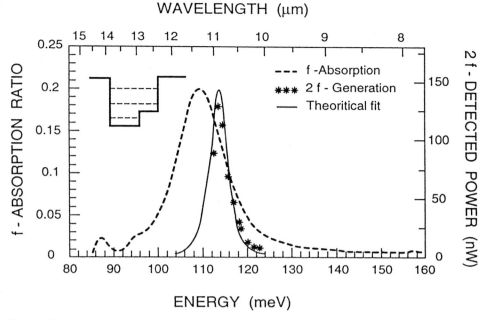

Figure 13 : Absorption spectrum (dashed line), second harmonic generated power (points) and theoretical fit (solid line) in a doubly-resonant quantum well structure. The SHG coefficient deduced from this measurement is 7. x 10^5 pm/V which is more than 3 orders of magnitude higher than in the bulk material [15].

structure, i. e. a structure in which $E_3 - E_2 = E_2 - E_1 = \hbar\Omega$, the second harmonic generation (SHG) term is given by[15] :

$$\chi^{(2)}_{2\omega} = \frac{q^3 (\rho_1 - \rho_2)}{\varepsilon_0 \hbar^2} \frac{\mu_{12} \mu_{23} \mu_{31}}{(\omega - \Omega - i\Gamma_2)(2\omega - 2\Omega - i\Gamma_2)} \qquad (14)$$

The large SHG coefficients obtained in AMQWs originate from i) the large dipolar moments μ_{ij} (as explained in the preceeding chapter), and ii) the small denominator $(\hbar\Gamma)^2$ at the double resonance condition $\omega = \Omega$. Figure 13 shows the variation of the SHG yield in an asymmetric MQW structure. The experimental value of the SHG coefficient is 7 x 10^5 pm/V, that is, more than 3 orders of magnitude higher than in the best nonlinear optical materials in this wavelength range[15] ! Recently, Capasso and his coworkers have realized such experiments with an applied electric field. They have been able to tune the QW resonant energies with the incident laser beam by linear Stark effect. This will allow to obtain electrically tunable optical nonlinear materials[16].

5. Conclusions

Enhancement of quantum effects in semiconducor quantum wells has led to a new generation of electro-optic devices. This enhancement is largely due to the small effective masses of the electrons in these materials and to the possibility to tune the optical resonances by band engineering. The field of application of these devices is mainly the optical computing and data processing (switching,...) as well as smart detectors. These properties should be even more pronounced when quantization will occur in other directions, in quantum wires and quantum boxes. Once technology will be mature, these latter will be real giant atoms exhibiting unprecedented electro-optic properties.

[1] G. Bastard, *Wave mechanics applied to semiconductor heterostructures* (Les Editions de Physique CNRS, Paris, 1988).
[2] R. Loudon, *The Quantum Theory of Light* (Clarendon, Oxford, 1983).
[3] C. Weisbuch and B. Vinter, *Quantized Semiconductor Structures: Physics and Applications* 1-252 (Academic Press, Boston, 1991).
[4] A. Yariv, *Quantum electronics* (John Wiley & sons, New York, 1989).
[5] K. L. Wang, S. K. Chun, and R. P. G. Karunasiri, in *Intersubband Transitions in Quantum Wells* E. Rosencher, B. Vinter, and B. Levine, eds. (Plenum, London, 1992) p. 227.
[6] B. F. Levine, in *Intersubband Transitions in Quantum Wells* E. Rosencher, B. Vinter, and B. Levine, eds. (Plenum, London, 1992) p. 43.
[7] S. Schmitt-Rink, D. S. Chemla, and D. A. B. Miller, Advances in Physics **38**, 89 (1989).
[8] C. Cohen-Tannoudji, B. Diu, and F. Laloë, *Mécanique Quantique* (Hermann, Paris, 1973).

9 C. Mead and L. Conway, *Introduction to VLSI sytems* (Addison Wesley, Reading,Massachusetts, 1980).
10 D. A. B. Miller, Optical and Quantum Electronics (1990).
11 L. C. West and S. J. Eglash, Appl. Phys. Lett. **46**, 1156 (1985).
12 B. F. Levine, C. G. Bethea, G. Hasnain, J. Walker, and R. J. Malik, Appl. Phys. Lett. **53**, 296 (1988).
13 E. Rosencher, E. Martinet, F. Luc, P. Bois, and E. Böckenhoff, Appl. Phys. Lett. **25**, 3255 (1991).
14 M. A. Kinch and A. Yariv, Appl. Phys. Lett. **55**, 2093 (1989).
15 E. Rosencher and P. Bois, Phys. Rev.B **44**, 11315 (1991).
16 F. Capasso, C. Sirtori, D. Sivco, and A. Y. Cho, in *Intersubband Transitions in Quantum Wells* E. Rosencher, B. Vinter, and B. Levine, eds. (Plenum, London, 1992) p. 141.
17 E. Martinet, F. Luc, E. Rosencher, P. Bois, and S. Delaître, Appl. Phys. Lett. **60**, 895 (1992).

High Speed Devices:
Properties and Applications

M.J. Kearney

GEC-Marconi Ltd., East Lane, Wembley, Middx. HA9 7PP, United Kingdom

1. Introduction.

Semiconductor devices form an integral part of electronic systems associated with the transmission and processing of information. The apparently insatiable requirements for faster information handling has therefore naturally meant a corresponding push for faster and faster semiconductor components and circuits. Current research efforts are directed towards realising high speed semiconductor devices whose intrinsic speed of response is considerably greater than 1 GHz in the case of analog signals, or 1 Gbit/s in the case of digital signals. Here we summarise these efforts and discuss the likely impact on present and future electronic systems.

The basic route to realising faster semiconductor devices is through miniaturisation. Crudely speaking, the two time scales which govern the upper speed of response of any device are a transit-time or a resistance-capacitance (RC) time. For a carrier to traverse a distance L takes a time of at least L/v_{max}, where v_{max} is the maximum carrier velocity. Decreasing L decreases this transit-time and to achieve intrinsic device speeds > 10 GHz requires length scales L < 1.5 µm (assuming a typical value of v_{max} ~ 10^7 cm s^{-1}), and usually shorter. The RC time constant is intimately associated with how long it takes to charge and discharge the device, and to reduce it requires the appropriate device lengths (in the case of resistance) and areas (in the case of capacitance) to be made smaller. The huge investment in Si VLSI (Very Large Scale Integration) based on standard Field Effect Transistor (FET) and bipolar transistor technology has seen the minimum device feature size (the gate length in a FET and the base width in a bipolar transistor) approximately halve every five years for the past twenty years, with commercially available circuits currently utilising ~ 0.8 µm gate lengths and < 0.1 µm base widths (translating into speeds of operation approaching 5 GHz or 5 Gbit/s for specialised digital logic circuits) and as many as 10^7 transistors. Besides increasing speed, such miniaturisation has also led to greatly enhanced reliability, reduced power consumption and reduced cost.

Despite the trend, there are clearly limits as to how far this process of miniaturisation can continue using conventional fabrication techniques. The resolution of lateral feature size is limited to ~ 0.1 µm using standard optical-lithography; X-ray lithography or electron-beam lithography is needed to fabricate features smaller than this. Using standard ion implantation or diffusion techniques to define accurate doping profiles over length scales ≤ 0.1 µm is difficult. Recent research on small scale, high speed semiconductor devices has owed much of its impetus to the advent of high precision epitaxial growth technologies such as Molecular Beam Epitaxy (MBE) and (slightly less precise) Metal-Organic Chemical Vapour Deposition (MOCVD). These allow for practically atomic layer growth control of semiconductors such as GaAs and, more recently, Si (a resolution of order 10 Å in the direction perpendicular to the substrate wafer). Simultaneously, there has been a refinement in heteroepitaxial growth, allowing semiconductor alloys such as $Al_xGa_{1-x}As$ and $Si_{1-x}Ge_x$ to be incorporated into GaAs and Si respectively. Since these alloys have different material properties depending upon their composition (most notably, different band-gaps), this has opened up entirely new possibilities in terms of device design and operation.

Continuing miniaturisation has important and non-trivial implications for the actual physics of device operation itself. For instance, in integrated circuits based on FET technology the typical voltages applied to the transistor must greatly exceed the thermal voltage kT/e (~ 26 mV at 300 K), otherwise thermal activation of carriers over internal potential barriers will prevent efficient switching action. Typically this means that one cannot reduce the working voltage below, say, 0.2 V, which means that as one continues to miniaturise, the internal electric field E (= V/L) will eventually have to increase. Effects relating to carrier heating by the field then have to be considered. For bipolar transistors, to maintain a working voltage V across the doped base requires a minimum doping density N ~ $2\varepsilon V/eL^2$ (where ε is the dielectric constant), otherwise punch-through occurs (total depletion). As base-widths become smaller, so the maximum practical levels of doping are approached. More fundamentally, if one can realise lengths scales comparable to the de-Broglie wavelength λ of free carriers in a semiconductor (λ ~ h/p, where h is Planks constant and p is the carrier momentum, and typically λ ~ 200 Å), then quantum confinement effects may become important. Understanding and even trying to utilise effects relating to the "physics of small length scales" is fast becoming an essential feature of high speed semiconductor device design.

These lectures are intended to provide an introduction to high speed devices and their applications. For detailed discussions about epitaxial growth and the more fundamental physics of novel semiconductor structures, the reader is referred to the other lectures found elsewhere in this volume.

2. Conventional semiconductor devices and their areas of application.

We begin our discussion by examining the relevant areas of application where speed of response is an important issue, foremost amongst which are digital electronics, microwave communications and optical-fibre communications [1]. Each of these areas is examined with emphasis on the limitations of some of the more important 'conventional' semiconductor devices (and materials) they utilise.

2.1 Digital electronics.

The handling of information in a digital representation offers significant advantages in terms of minimising unwanted errors arising from system imperfections. Digital logic, implemented in Si VLSI with Si transistors providing switches for processor and memory elements, has heralded one of the great technological revolutions of the 20th century. There are several reasons why Si is the most widespread semiconductor used for digital logic; Si is cheap and abundant, it is structurally robust, it can be readily doped either n-type or p-type and, from an environmental point of view, it is relatively harmless. Since it was one of the first semiconductors studied in detail, the cumulative investment of time and money has led to extremely sophisticated and refined processing and fabrication facilities. The fact that Si has a natural and very high quality dielectric (insulating) oxide, namely SiO_2, has been instrumental in giving rise to digital integrated circuits (ICs) based on MOS (Metal Oxide Semiconductor) FET technology, particularly CMOS (C standing for complementary) which makes intricate use of both n-type and p-type MOSFETs in a single circuit to minimise problems associated with power dissipation.

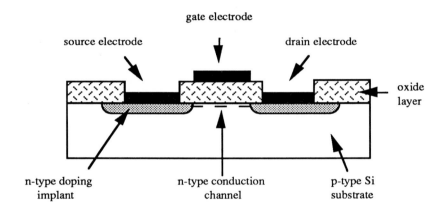

Fig.1. Schematic representation of an n-channel Si MOSFET.

In Fig. 1 we show a schematic of the basic n-channel MOSFET structure. By applying a voltage V_{gs} to the gate one can control the source-drain current I_{ds} by altering the number of charge carriers beneath the gate. The defining characteristic of any transistor is its gain (a small change in one parameter resulting in a large change in another); in this case a small change in the source-gate voltage V_{gs} results in a large change in the source-drain current I_{ds}. The intrinsic transconductance is a measure of this gain and is defined by,

$$g_{mi} = \left.\frac{\partial I_{ds}}{\partial V_{gs}}\right|_{V_{ds}} \approx \frac{C_g}{\tau} \tag{1}$$

where C_g is the gate capacitance and τ is the average carrier transit-time between the source and the drain. The variable τ is clearly a measure of how quickly the source-drain current can respond to a change in gate-voltage, and a figure of merit for a transistor is the frequency $f_T = 1/2\pi\tau$ at which the current gain (as deduced from network analyser measurements under conditions where the output is short-circuited) falls to unity. The cut-off frequency f_T is one important parameter which governs the fundamental speed of digital logic circuits. Since we can approximate τ by L/v, where L is the effective transit length under the gate and v is the average carrier velocity during transit, a high gain and a high value of f_T requires small gate lengths and high carrier velocities.

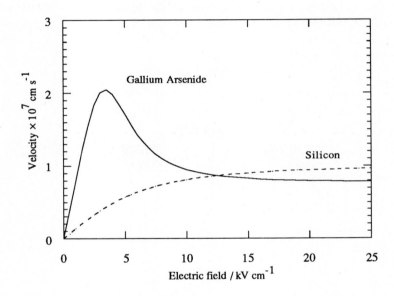

Fig. 2. Velocity-field characteristics for GaAs and Si.

As far as individual transistor action is concerned, the electrical properties of Si are not especially good. In Fig. 2 we compare the *equilibrium* velocity-field characteristics of n-type Si and n-type GaAs [2]. At very low fields the velocity (v) is linearly related to the field (F) by the mobility μ ($v = \mu F$), and for GaAs μ is higher (up to 8,000 $cm^2V^{-1}s^{-1}$ at 300 K in pure material as opposed to typically 1,500 $cm^2V^{-1}s^{-1}$ in Si), due to the small Γ-valley electron mass in GaAs. The higher value of μ translates into significantly higher velocities at low fields. For fields greater than ~ 3 kV cm^{-1}, however, the velocity of electrons in GaAs begins to decrease as they are transferred (heated by the field) into the higher energy L and X satellite valleys which have larger electron masses. The bandstructure of Si is such that no analogous effect occurs in its case. Eventually, as the field is increased further, the velocity in both Si and GaAs saturates to a value which is similar for both. The equilibrium v(F) relationship may be used to estimate τ for a given source-drain field in FETs with gate lengths > 1 μm, but for FETs with gate lengths ≤ 0.5 μm the question of how quickly carriers move between the source and drain is more subtle. In particular, carrier transfer may be so rapid (< 3 ps) that (in GaAs) carriers have insufficient time to transfer to the satellite valleys whereupon, because of the low Γ-valley mass, significant *velocity overshoot* can occur (this effect is much less prevalent in Si). The generally higher carrier velocities and this overshoot phenomenon give n-type GaAs a distinct advantage in basic speed over Si, and for very high-speed digital logic GaAs ICs are routinely employed. The lack of a suitable dielectric analogous to the oxide in Si is a drawback and there is no equivalent GaAs MOSFET technology. At the single device level the n-type MESFET (Metal-Semiconductor FET, where the gate consists of a metal-semiconductor Schottky contact with a typical barrier height of ~ 0.9 V, see § 4) is a more than adequate replacement, with a higher value of f_T for a comparable gate length, but the degree of GaAs MESFET circuit integration is not yet comparable with that of Si MOSFETs. Further, the somewhat poorer transport characteristics of holes in p-type GaAs have hampered the development of an effective complementary technology.

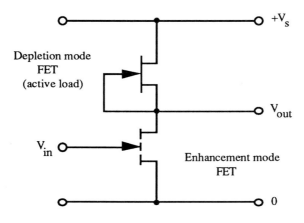

Fig. 3. A two FET inverter gate with active load.

As far as digital logic is concerned, the ultimate performance of a circuit depends upon more than just the source-drain transit-time (or f_T value) of an individual transistor; how the circuit behaves as a whole is equally important. The basic building block of all logic circuits is an inverter, an example of which is shown in Fig. 3 and which forms the basis of so-called direct coupled FET logic (DCFL). This particular inverter consists of two FETs, an enhancement-mode FET (designed not to conduct when $V_{gs} = 0$) and a depletion-mode FET (which does conduct when $V_{gs} = 0$). The inverting action of this logic gate is clear; a high value of V_{in} results in a low value of V_{out} and vice versa. The dynamic switching speed of an inverter, usually measured by studying the frequency at which the gain of a ring oscillator consisting of an odd number of such inverters becomes unity, defines a propagation-delay time τ_D. This is greater than the source-drain transit-time because now the charging-times for various capacitances are also taken into account. For a ring-oscillator, each inverter is connected to one other (a fan-out of one), and in more complicated circuits (where the fan-out is usually greater), the overall speed of response will be slower still (note that the clock rate of a 486 microprocessor is 'only' 33 MHz). Related to τ_D is the power-delay product $P_D\tau_D$, where P_D is the average switching power, which defines the typical energy required to make one logic transition. The most important criterion for digital circuits is to obtain the highest possible switching speed for the lowest overall dynamic power consumption.

It is interesting to reflect on the various merits of Si CMOS versus n-type GaAs MESFET digital logic. The former allows for much higher levels of integration, offers a lower *static* power consumption and is invariably cheaper and easier to implement. In simpler circuits (LSI and smaller) the latter generally offers an advantage in speed, which can only be matched by operating Si CMOS at higher supply voltages, a process which raises the *dynamic* power consumption to comparable levels. In fact, for very high speed (high clock rate) circuits GaAs DCFL offers an overall power consumption which is usually somewhat smaller than that of all Si based logic technologies. At the very highest levels of integration the difference in speed attributed to individual components becomes less significant and the performance of the two technologies at the VLSI level appears comparable [1].

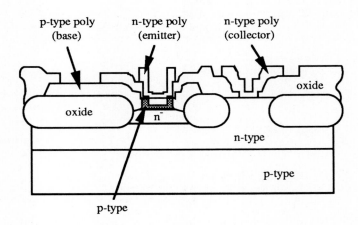

Fig. 4. Schematic of a modern Si n-p-n bipolar transistor structure.

Besides the Si MOSFET and GaAs MESFET, the other major transistor in widespread use today in digital logic is the Si bipolar transistor. A schematic of an n-p-n device (the complementary p-n-p structure is constructed in a similar way) is shown in Fig. 4, and here the transistor action arises when a voltage applied to the base lowers the potential barrier at the base-emitter p-n junction, allowing electrons to flow from the emitter to the collector. The emitter-collector transit-time determines the value of f_T, much as the source-drain transit-time determines f_T in a FET, and since it is generally easier to control a base thickness through ion implantation or diffusion than a gate length through optical lithography, Si bipolar devices usually have higher f_T values than Si MOSFETs and smaller propagation delay times (as long as the emitter width is made small as well), and are therefore favoured for very high speed ICs. The exponential increase in output current with base voltage also leads to a much higher transconductance g_m and therefore gain, and since the entire emitter area conducts (unlike only a narrow channel in a FET), bipolar transistors can switch large current densities. The turn-on voltage V_{be} corresponds to that of a p-n junction and can be made much more uniform across an IC than the turn-on voltage of Si MOS and GaAs MESFET ICs, where the metal-semiconductor barrier heights determining V_{gs} are much more prone to variation. This, coupled with the high value of g_m, enables the voltage swing in logic operations to be reduced (for a given bit-error rate) which helps to minimise the power-delay product. In one common logic implementation, so-called emitter-coupled logic (or ECL), the power dissipated per gate is almost independent of the gate clock rate; this should be compared with Si MOS technology where there is a substantial difference between the static and the dynamic power consumption.

Table 1. Representative performance for conventional logic transistors

	Si MOSFET	GaAs MESFET	Si Bipolar
gate length	0.8 μm	0.8 μm	
emitter width			0.7 μm
base thickness			0.1 μm
f_T	5 GHz	20 GHz	20 GHz
τ_D (ring oscillator)	100 ps	30 ps	50 ps
$P_D \tau_D$	10 fJ/gate	30 fJ/gate	50 fJ/gate

Bipolar technology does have some disadvantages compared to FET technology. Only enhancement-mode bipolar devices are available. FETs are unipolar devices and are not affected by minority carrier storage, something that eventually degrades the speed of bipolar devices. A temperature rise due to excess heat dissipation within a bipolar transistor can lead to an increase in collector current, so further increasing the heat dissipated and potentially leading to thermal runaway. The input impedance of FETs is much higher (which simplifies the design of circuits with a large number of stages) and they require essentially only a voltage to switch them; the low input impedance of bipolar transistors means that a finite base current must be drawn which significantly increases their static and dynamic power dissipation. The variation

$g_m(V_{gs})$ is closer to being linear in a FET than that of $g_m(V_{be})$ in a bipolar transistor; this reduces distortion in certain analog applications (see below). Finally, Si Bipolar VLSI circuits are more difficult to fabricate than Si CMOS VLSI circuits (compare Fig. 1 and Fig. 4). The typical performance parameters for commercially available logic transistors are shown in Table 1. A lot of recent effort has been devoted to BICMOS technology, which tries to combine the best features of both types of Si device in one circuit; namely the high transconductance of the bipolar transistors and the high input impedance of the FETs.

2.2 Microwave applications.

Besides digital logic, the transistors described above are widely used in analog circuits such as digital-to-analog (D/A) and analog-to-digital (A/D) converters, small signal amplifiers, power amplifiers and oscillators. The detection and generation of microwave radiation at frequencies above ~ 1 GHz requires specialist high speed amplifier and oscillator circuits, but such a push to higher frequencies is an inevitable consequence of the ever growing congestion in the lower (more readily accessible) regions of the spectrum. For communications purposes, the higher the carrier frequency, the greater the potential bandwidth for high data rate transfer as well. Current areas of application include line-of-sight links for local area networks (LANs), mobile communications and direct broadcast satellite receivers, presently operating at frequencies up to about 10 GHz. In the millimetre-wave part of the spectrum (30 GHz corresponds to a radiation wavelength of about 1 cm), particular frequencies of interest are 35 GHz, 94 GHz, 140 GHz and 220 GHz where the atmospheric attenuation has local minima (a requirement for long distance transmission), and 60 GHz where the atmospheric attenuation is very high (useful for local transmission without causing long range interference). As well as communications, radar systems operating at these frequencies are also being pursued in a commercial (as well as military) context; for example, a great deal of research is currently being devoted to realising cheap collision avoidance radars operating at 60 - 70 GHz for cars fitted with intelligent cruise control.

Amplifiers are an integral part of microwave receivers. In a transistor amplifier, the inherent gain A of the transistor is used to convert an input signal V_{in} to an output $V_{out} = AV_{in}$. In practice, a fraction F of the output is fed-back to the input; thus $V_{out} = A(V_{in} + FV_{out})$ or $V_{out} = AV_{in}/(1-AF)$. If the loop-gain AF is negative (negative feedback), the overall gain of the amplifier is reduced but now the output is stabilised against fluctuations in the input. To reduce distortion it is desirable that the transistor should be as linear as possible; thus FETs are attractive for small-signal amplifiers because their variation of g_{mi} with input voltage is less than that of bipolar transistors (see § 2.1). If the loop-gain of an amplifier is made positive and such that AF > 1, then the combination of transistor power gain plus feedback produces a circuit instability (negative terminal resistance) which, when coupled to a suitable resonating element (an equivalent LC circuit), can lead to stable oscillations. This forms the basis of a microwave source. The non-linearities of the transistor prevent the instability growing without limit and it should be noted that, unlike the amplifier, oscillator circuits are inherently non-linear.

An important figure of merit for transistors utilised in microwave amplifiers and oscillators is f_{max} (the maximum frequency of oscillation), which is the frequency at which the maximum available power gain of the transistor becomes unity. This frequency is not necessarily the same as f_T but is related to it by impedance ratios; thus for FETs,

$$f_{max} \sim \frac{f_T}{2\left[G_{ds}(R_g + R_s) + 2\pi f_T C_g R_g\right]^{1/2}} \qquad (2)$$

where G_{ds} is the source-drain conductance, R_s is the source resistance and R_g is the gate resistance. A similar expression can be written down for bipolar transistors (usually with the first term in the denominator neglected) with base replacing gate. Depending upon the device in question f_{max} may be greater than or less than f_T, and a large value of f_{max} is favoured by having a large value of f_T coupled with low parasitic resistances. Just as f_T does not necessarily coincide with the frequency of operation of digital circuits, so f_{max} does not necessarily represent the maximum frequency of operation of an oscillator or amplifier circuit. Values of f_{max} approaching 200 GHz have been reported for GaAs MESFETs with gate lengths ≤ 0.25 μm [1].

Another important consideration for microwave circuits is the noise added to the output by the transistor. The signal-to-noise ratio at the input divided by the signal-to-noise ratio at the output defines the transistor noise figure NF, and a common expression for the noise figure of a FET is,

$$NF = 1 + \frac{f}{f_T} K C_g \left((R_s + R_g) g_{mi}\right)^{1/2} \qquad (3)$$

where K is an empirical (Fukui) fitting parameter [3] and f is the frequency. This expression assumes that the dominant noise source is thermal (or Johnson) noise arising in the parasitic resistances R_s and R_g due to the random Brownian motion of the charge carriers. At low frequencies, however, 1/f or flicker noise also becomes significant, which is believed to be due to the random filling and emptying of traps at the metal-semiconductor or oxide-semiconductor interfaces and at the 'interface' of the conduction channel and the resistive substrate, although its origins are still not fully understood. For small signal amplifiers operating well above the corner frequency (i.e. the frequency below which the 1/f noise begins to exceed the thermal noise, which is typically between 10 and 100 MHz for a FET), 1/f noise can usually be ignored [4]. Typical performance for state-of-the-art commercially available GaAs MESFETs are noise figures of order 1 dB at 10 GHz, with an associated gain of $\geq$ 10 dB (dB stands for decibel and to convert to dB take $10\log_{10}$ of the relevant quantity). For oscillators, where the non-linearity can result in substantial up-conversion of 1/f noise to frequencies near the fundamental oscillator frequency, this is not the case. The absence of equivalent trapping interfaces means that the 1/f noise of bipolar transistors is usually much lower than FETs, which is advantageous for oscillator applications.

GaAs MESFETs are presently by far and away the most important microwave transistor [5]. The importance of GaAs to the microwave industry stems not just from its higher carrier velocity (compared to Si), but from the availability of high quality semi-insulating GaAs substrates (a two orders of magnitude improvement in resistivity over Si to ~ 10^8 Ωcm). This aids in the fabrication of Monolithic Microwave Integrated Circuits (MMICs) by reducing substrate leakage and allowing for the fabrication of low-loss transmission lines to connect the various circuit elements together (transmission lines are required once the wavelength of the radiation becomes comparable to the circuit size). The advantages inherent in integrating as many microwave components as possible on a single chip include lower cost, reduced parasitics and more reproducible performance. For frequencies of operation below about 30 GHz one can readily fabricate MMIC receivers with all the mixer and detector diodes (see below), amplifier circuits and transmission lines defined on a single substrate. Often, however, full integration is not possible and hybrid circuits have to be used, consisting of an integrated circuit plus a few discrete components.

The mixer and detector diodes referred to above are used to detect return signals in the case of radar and to demodulate the information encoded onto a carrier frequency in the case of communications. Suppose we have an amplitude modulated signal of the form,

$$V_{sig} = A(1 + m\cos 2\pi pt)\cos 2\pi ft \tag{4}$$

where f is the carrier frequency and p is the modulation frequency (we only need to consider a single modulation frequency since one can resolve any modulation into discrete Fourier components, each of which may be analysed separately). By re-writing Eq. (4) as follows,

$$V_{sig} = A\cos 2\pi ft + \frac{1}{2}mA\cos 2\pi(f+p)t + \frac{1}{2}mA\cos 2\pi(f-p)t \tag{5}$$

we see that the signal consists of the carrier frequency plus side-bands at frequencies f ± p. The more information that is required to be transmitted, the greater the required spread of modulation frequencies or bandwidth. To extract the modulation frequency one cannot simply filter out the carrier frequency f. The simplest way to demodulate the signal is to first pass it across a rectifying diode which has a non-linear current-voltage (I-V) characteristic. If we expand the I-V characteristic as a Taylor series,

$$I = I_0 + aV_{sig} + bV_{sig}^2 + ...$$

$$I = I_0' + bA^2 m\cos 2\pi pt + \text{higher frequency terms} \tag{6}$$

we see that the rectifying action of the diode now allows the modulation frequency to be separated and the higher frequencies to be removed using a suitable low-pass filter (hence one usually requires the condition p << f, which means that a high carrier frequency is needed to support a large modulation bandwidth). The most commonly used detector diode is the metal-semiconductor Schottky diode, which has an I-V characteristic similar to that of a p-n junction (see § 4), although p-n junctions themselves tend not to be used because of high-speed limitations associated with minority carrier injection. Ideally one would like to amplify the signal before detection so that the diode is driven into the regime where its forward-bias I-V characteristic saturates and becomes linear, which significantly improves conversion efficiency and reduces harmonic distortion compared to the small-signal square-law detection scheme outlined above [4].

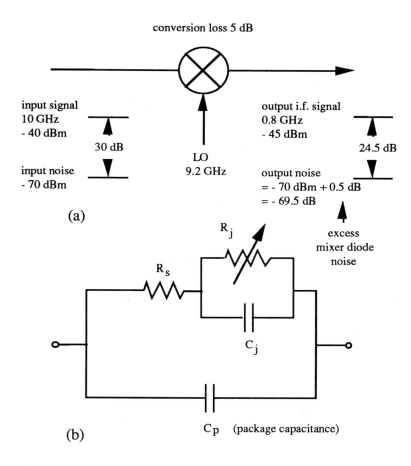

Fig. 5. (a) Schematic of the mixing process with plausible parameter values, (b) equivalent circuit for a Schottky diode.

The difficulties associated with high frequency amplification, however, means instead that it is usual to first mix the signal with that of another source (a local oscillator or LO) in a non-linear Schottky mixer diode to generate a new modulated signal but on a carrier which is at a lower frequency (known as the i.f. or intermediate frequency), the difference between the original carrier frequency and that of the LO [4].

The down-converted signal can then be amplified in a low-noise i.f. amplifier before being detected, a process known as superheterodyne detection (see Fig. 5). To reduce the effect of 1/f noise the i.f. frequency is usually set to be well above the corner frequency. Both Si and GaAs Schottky diodes are used as mixers and detectors, although a GaAs Schottky diode requires essentially the same fabrication steps as a GaAs MESFET, and so lends itself naturally to monolithic integration.

The tangential sensitivity is a measure of the weakest signal a detector diode can reliably detect against a noise background (corresponding to a signal-to-noise ratio of about 2.5), and at 10 GHz a good figure is ~ - 55 dBm (dBm means power measured relative to 1 mW). The tangential sensitivity for a given diode may be shown to degrade with frequency (deduced from the equivalent circuit in Fig. 5) as $1 + (f/f_c)^2$, where the cut-off frequency $f_c \sim 1/(2\pi C_j (R_s R_j)^{1/2})$, with C_j and R_j the junction capacitance and resistance respectively, and R_s the parasitic series resistance [4]. To achieve a high level of performance one must satisfy the criterion $f \ll f_c$, but the cut-off frequencies of Schottky diodes can be as high as THz. For a mixer, the available power at the signal frequency divided by the available power at the i.f. frequency defines the conversion loss, and related to this is the noise figure (the ratio of input to output signal-to-noise ratios, which includes the excess diode noise contribution, see Fig. 5). The overall noise figure (ONF) includes the noise contribution from the i.f. amplifier as well. The conversion loss of a mixer diode whose LO drive level is optimised for a given frequency scales with frequency as $\sim 1 + 2f/f_c$, where $f_c \sim 1/(2\pi C_j R_s)$. To reduce the conversion loss and ONF as much as possible requires the implementation of quite sophisticated circuit techniques (such as balanced mixing for example), as well as reducing parasitics. For a mixer at 10 GHz, the conversion loss of a good GaAs Schottky diode is about 4.5 dB.

Unlike a microwave receiver, the full integration of an oscillator power source is more difficult since providing a suitable resonating element at high frequencies is non-trivial. Further, the maximum frequency at which significant quantities of power ($\geq$ 100 mW) have been generated by a GaAs MESFET (in conjunction with a dielectric resonator, a so-called DRO) is about 30 GHz [5]. Above 30 GHz, the performance of GaAs MESFET oscillators begins to fall markedly. The most important power sources in the millimetre-wave regime are the Gunn diode and the IMPATT (Impact Avalanche Transit-Time) diode, both of which utilise a cavity as a resonating element which must be individually (mechanically) tuned for each diode (as a result of which these sources are usually expensive). The Gunn diode achieves the required negative terminal (or differential) resistance (NDR) by exploiting the negative differential mobility of GaAs at fields above ~ 3 kV cm^{-1} (see Fig. 2). Si, which does not exhibit this effect, therefore cannot be used as a Gunn diode material. The basic Gunn diode structure is shown in Fig. 6. Under an applied bias sufficient to ensure that the threshold-field of 3 kV cm^{-1} is exceeded in the drift region (equally referred to as the transit region), a steady-current is unstable and is replaced instead by discrete charge domains which form at the cathode and propagate towards the anode at the saturated drift velocity v_s. Eventually these charge domains dissipate and the process begins again. This gives rise to periodic voltage and current variations in the circuit [2,4] with the current pulses occurring out of phase with the voltage (Fig. 7), which is the usual manifestation of NDR.

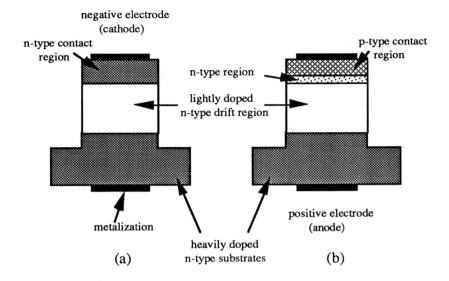

Fig. 6. (a) Gunn diode, (b) single-drift Hi-Lo IMPATT diode.

The (fundamental) frequency f of the oscillations is intimately connected with the average domain propagation time, and if the domains reach the anode before dissipating, then $f \sim v_s/L$ (where L is the length of the drift region). High frequency devices therefore require short transit region lengths (i.e. $L < 2$ μm for $f > 50$ GHz). Based on the equivalent circuit depicted in Fig. 7 we note that if the average voltage over one cycle is V_{dc} and the average current I_{dc}, then the power dissipated in the diode is $P_{dc} = V_{dc}I_{dc}$. The RF power supplied to the load (assuming an idealised sinusoidal voltage) is given by,

$$P_{rf} = \frac{V_{rf}}{T} \int_0^T I(t)\sin(2\pi ft)dt \qquad (7)$$

and the efficiency of DC to RF power conversion is simply the ratio $\eta = P_{rf}/P_{dc}$. Standard GaAs Gunn diodes are capable of delivering about 200 mW of power continuously (i.e. not pulsed) at 35 GHz (with ~ 8 % efficiency) falling to 20 mW (2 % efficient) at 94 GHz (operating in second harmonic mode). The large quantities of dissipated power and high current densities ($\geq 10^5$ A cm^{-2}) require that great attention is paid to providing adequate heat-sinking (through the incorporation of diamond heat-sinks for example). The III-V material InP may also be used to make Gunn diodes, and because the threshold field for NDR in InP (~ 10.5 kV cm^{-1}) is higher than that in GaAs, translating into higher bias voltages, InP devices are usually more powerful. A drawback, however, is that InP as a substrate material is not particularly robust and needs to be handled with care.

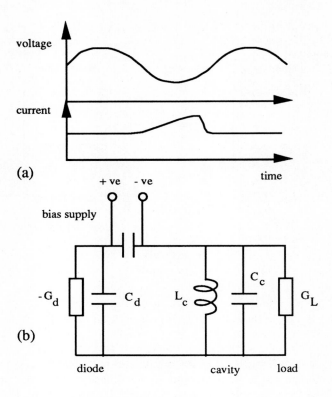

Fig. 7. (a) Idealised current and voltage waveforms for a diode exhibiting NDR, (b) lumped-element equivalent circuit for the diode and its cavity.

IMPATT diodes are much more powerful and efficient than Gunn diodes, but also much more noisy and prone to catastrophic failure. Unlike the Gunn diode, Si can be used as an IMPATT material and for frequencies of operation greater than about 50 GHz, the most powerful sources tend to be Si double-drift IMPATT diodes. For frequencies less than 50 GHz, GaAs double-drift and single-drift structures (similar to that depicted in Fig. 6) are preferred. In an IMPATT diode the device is reverse biased to the point where avalanche breakdown can occur. Avalanche electrons generated at the p-n junction (with reference to Fig. 6) propagate across the drift region to the anode, and the combination of this transit-time delay plus the inherent delay due to the time it takes the avalanche current to build up to its maximum value (~ half a period) leads to NDR. To maximise the NDR one sets the transit region length L according to $f \sim v_s/2L$. The large breakdown voltages account for the large amounts of generated power, while the high efficiencies result from the fact that the current can be made more 'out-of-phase' with the voltage (Fig. 7) than in a Gunn diode (the current tends to be small during the positive portion of the voltage cycle where the voltage falls below the threshold for avalanching). State-of-the-art output powers (efficiencies) for GaAs double-drift structures include 10 W (20 %) at 10 GHz and > 1 W (13 %) at 60

GHz [6]. The poor noise properties of IMPATT diodes are intimately associated with the statistics of the avalanching process (discussed in more detail in the context of avalanche photodiodes in the next section).

2.3 Optical-fibre communications.

Optical-fibre links (utilising carrier frequencies close to those of visible light) have had a very significant impact on the telecommunications industry, and one can envisage that house-hold fibre links will soon become as commonplace as conventional telephone lines. The principal attraction of optical-fibre communications is the high carrier frequency (close to 10^{15} Hz), which potentially allows an enormous amount of information (high bandwidth) to be carried on just a single line. Light travelling down a fibre is free from the effects of atmospheric distortion which plague microwave communication systems, although the inevitable frequency dependent dispersion and attenuation within the fibre are significant design criteria for longer distance telecommunications purposes. For short distances, however, these effects are less deleterious. At the shortest scales, it has been proposed that optical-fibre interconnects may soon help circumvent the traditional problem of providing non-cumbersome high data rate electrical connections between individual chips, circuit boards and circuit board racks in digital electronic systems.

An optical-fibre communications link requires (in addition to a fibre), a light source which can be rapidly modulated and a detector (receiver) which can demodulate and interpret the signal. For short distances (up to a few km) a Light Emitting Diode (LED) will do as a source, but usually a semiconductor laser diode is preferred since the available power is greater, that power can be more rapidly modulated and the tight, coherent beam is more easily coupled into a fibre. Commonly used detectors are PIN diodes and avalanche photodiodes (see below). Depending upon the required wavelength of operation, different semiconductor materials tend to be used for these devices. A fundamental limitation of Si as a semiconductor is the fact that, unlike the III-V semiconductors GaAs and InP, its band-gap is *indirect*. This makes it unsuitable for most optoelectronic applications, although Si can absorb light and is widely used in solar cells for example. The direct band-gap of GaAs, InP and related III-V materials is required to form an efficient radiation source.

If one defines the cut-off wavelength of a semiconductor by $\lambda_c = hc/E_g$, where E_g is the band-gap and c is the velocity of light, then simplistically λ_c defines the typical wavelength of radiation that can be readily absorbed or emitted. For GaAs ($E_g = 1.42$ eV) and InP ($E_g = 1.35$ eV) at 300 K, $\lambda_c \sim 0.9$ μm (near infrared). The human eye is sensitive to wavelengths between about 0.77 μm (red) and 0.39 μm (violet) and so for display applications neither GaAs or InP is adequate. Instead, the III-V ternary alloy $GaAs_{1-x}P_x$ is commonly used, with which λ_c can be extended through the red and into the yellow. The search for a cheap, bright and reliable blue light source remains elusive. As far fibre communications are concerned, for short distances GaAs LEDs and lasers are used as sources and the signals they send may be readily demodulated using a Si PIN photodetector or avalanche photodiode (for Si, $E_g = 1.12$ eV and $\lambda_c \sim 1.1$ μm). For longer distance transmission one ideally wants to work at wavelengths corresponding to the low loss attenuation regions of silica optical-fibres, namely 1.3 μm for which the attenuation is ~ 0.6 dB/km (and the chromatic dispersion is a minimum), and 1.55 μm for which the attenuation is ~ 0.2 dB/km (for comparison, at 0.9 μm the typical attenuation is 1.5 dB/km). Here the III-V quaternary compounds $Ga_xIn_{1-x}As_yP_{1-y}$ are particularly useful which, for suitable choices of x and y, have the same lattice constant as InP (lattice matched) and can be grown without strain being present on InP substrates. The appropriate values of λ_c span the range 0.9 μm (InP) to 1.67 μm ($In_{0.53}Ga_{0.47}As$), depending upon precise composition. For wavelengths greater than 0.9 μm, InP is transparent and so the emitted light can be transmitted through the substrate, which makes device fabrication simpler.

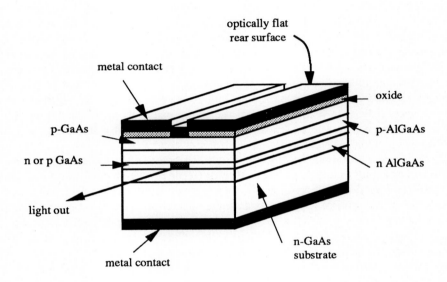

Fig. 8. Schematic of a stripe contact GaAs/AlGaAs double heterostructure laser.

An LED is basically a forward biased p-n junction, where electrons and holes from the heavily doped n-type and p-type layers combine and in doing so emit photons whose energy corresponds roughly to that of the semiconductor band-gap. In a laser, these photons bounce backward and forward (positive feedback) between two optically flat surfaces which act as mirrors, encouraging stimulated emission to occur and the radiation amplitude to increase (gain) to the point where light can penetrate the weaker mirror in a tight, coherent beam. Conventional homojunction devices of the type just described tend to require very large current densities to achieve the threshold for lasing (J_{th} ~ 5.0 x 10^4 A cm^{-2}), which leads to formidable problems in terms of excess heat dissipation. The preferred double heterostructure (DH) laser depicted in Fig. 8 exploits the larger band-gap of $Al_xGa_{1-x}As$ (with an Al mole fraction typically of the order 0.3) to confine the electrons and holes within the active region, which improves the electron-hole overlap and hence improves the recombination efficiency. This significantly reduces the value of J_{th} to a value typically of order 10^3 A cm^{-2}. As a further benefit, the light is also confined and waveguided by the dielectric mismatch at the GaAs/Al$_x$Ga$_{1-x}$As interfaces which improves stimulated emission efficiency [2].

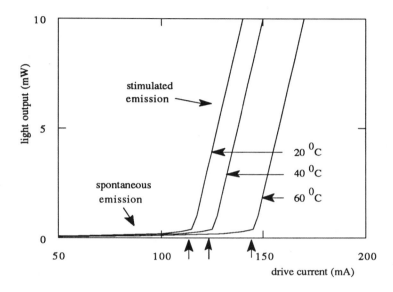

Fig. 9. The output power of a typical 20 μm stripe-width DH laser diode showing the sharp onset at the threshold current (marked with arrows). Notice the temperature sensitivity of the threshold current.

Modulation is conventionally achieved by directly modulating the current which drives the laser diode; the intensity of the output light is a strong function of current once threshold is exceeded (see Fig. 9). Most optical-fibre communication schemes utilise digital as opposed to analog modulation for two reasons; firstly, relatively small current swings near threshold are sufficient to provide a high on-to-off contrast ratio and secondly, most commonly used detectors respond directly to light intensity rather than amplitude [7]. When subjected to a sinusoidal modulation of frequency f, the response of LEDs and detectors (strictly speaking the relevant expression for a laser is more complicated) can usually be represented in the following way [1,2,7],

$$R(f) = \frac{R(0)}{\left(1 + 4\pi^2 f^2 \tau^2\right)^{1/2}} \quad (8)$$

where τ is an effective carrier 'lifetime'. The 3 dB (or modulation) bandwidth is defined by $f_{3dB} = 1/2\pi\tau$. The ultimate modulation speed is governed by intrinsic properties associated with the laser, such as the effective carrier lifetimes before recombination and the differential gain, as well the capabilities of the driving circuit. Modulation bandwidths of order 20 GHz have been demonstrated with DH laser diodes, well above the standard data transfer rates currently utilised in fibre communications (typically less than 565 Mbit/s; one telephone line utilising 64 kbit/s corresponds to a bandwidth requirement of 4 kHz).

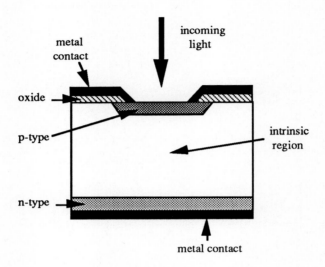

Fig. 10. The cross-section of a typical Si PIN photodiode.

A commonly used photodetector in the wavelength range 0.4 μm to 1 μm is the Si PIN diode (Fig. 10). During operation the device is reverse biased so that the electric field in the intrinsic region separating the p and n-type semiconductor regions is large. Light entering this intrinsic transit region creates electron-hole pairs which are swept-out by the field and collected in the n and p-type regions, causing an external circuit current to flow which can be detected. The quantum efficiency (the number of electron-hole pairs generated per incident photon) is given by [2],

$$\eta = (1-R)\left(1 - \frac{e^{-\alpha L}}{1+\alpha L_p}\right) \quad (9)$$

where R is the surface reflection coefficient (which can be made very small by the use of an anti-reflection coating), $\alpha(\lambda)$ is the absorption coefficient for light of wavelength λ, L is the drift region thickness and L_p is a hole diffusion-recombination length. In a carefully designed structure the modulation bandwidth of a photodiode is usually dominated by transit-time effects and is given by [2],

$$f_{3dB} \approx \frac{0.4 v_s}{L} \qquad (10)$$

To improve the modulation bandwidth one should make L small; however, this also reduces the amount of light absorbed and hence the quantum efficiency, and the optimum trade-off between speed and efficiency occurs when L ~ 1.5/α. For Si photodiodes operating at λ ~ 0.8 μm, α ~ 1000 cm^{-1} and so L ~ 15 μm. Such devices can exhibit quantum efficiencies of order 75 % with a simultaneous modulation bandwidth of order 2.5 GHz. Notice that if the absorption coefficient is too high (> 10^4 cm^{-1}) or the surface layers too thick, then a significant amount of light will be absorbed in surface regions where the recombination-time is very short, reducing efficiency substantially.

The rms modulated input power required to produce an output signal-to-noise ratio of 1 in a bandwidth of 1 Hz is referred to as the noise equivalent power or NEP. For a PIN diode we have [2],

$$NEP = \sqrt{2} \frac{\left(q I_B + q I_D + 2kT/R_{eq}\right)^{1/2}}{R_\lambda} \qquad (11)$$

where I_B is the current arising from background radiation, I_D is the thermally generated dark current and R_λ is the responsivity (output current per unit input signal power). This expression accounts for the fact that both the background and dark currents have associated shot noise and there is also thermal noise generated in the equivalent external circuit resistance. To improve sensitivity, one needs to improve R_λ and reduce I_B and I_D. In an avalanche photodiode (APD), the basic PIN diode structure is reverse biased to the point where the electric field is sufficient to cause avalanche multiplication to take place. Electron-hole pairs created in the transit region then create other electron-hole pairs leading to rapid carrier multiplication and internal current gain. By this means the responsivity of the diode can be significantly increased, but the shot noise is also amplified and this needs to be considered carefully. In particular, the shot noise current does not simply increase as the average carrier multiplication factor M, but there is also an additional effect due to the random statistical nature of the multiplication process itself, expressed in terms of the excess noise factor F(M). Under the assumption that electrons are the dominant carrier type initiating multiplication in the first place one may show that [8],

$$F(M) = M\left[1 + \left(\frac{\alpha_p}{\alpha_n} - 1\right)\left(\frac{M-1}{M}\right)^2\right] \qquad (12)$$

where α_p and α_n are the hole and electron ionisation rates respectively (these terms are interchanged if holes are the dominant carrier type initiating multiplication. For GaAs α_p and α_n are roughly equal and so F(M) ~ M. In Si, however, the ratio α_n/α_p is strongly field dependent and may be as large as 30, which significantly reduces F(M) according to Eq. (12). Careful optimisation of the doping profile has led to Si APDs operating in the range 0.6 μm to 1.0 μm which exhibit quantum efficiencies ~

100 % and excess noise factors F(M) as low as 4 for values of M ~ 100 [2]. (For wavelengths greater than 1 µm Ge APDs are often used, although a problem here is that narrower band-gap semiconductors naturally have a correspondingly larger dark current). A high gain and low value of F(M) is favourable in so much as it reduces the importance of the thermal noise in the equivalent external circuit resistance, thus improving the signal-to-noise ratio, which is particularly useful when the signal to be detected is weak. However, a high gain also implies a correspondingly long avalanche process and hence a reduced bandwidth; the gain-bandwidth product is usually roughly constant and is the commonly quoted figure of merit for APDs.

The growing importance of fibre communications has led to considerable research on realising Optoelectronic Integrated Circuits (OEICs), whereby the various sources, detectors, interconnects and associated signal processing electronics are defined on a single circuit or chip. This is clearly most easily accomplished in the case of Si photodetectors and GaAs light sources, but much effort has also been expended on defining high speed electronic circuitry in InP based materials. The potential advantages inherent in integrated optics, that is in carrying out as much of the signal processing as possible optically without first converting into electronic signals (using a variety of thin film planar structures) has led to much interest both in the context of more efficient communications and also ultra-high speed optical computing. To date, however, this latter concept of integrated optics has had only limited commercial impact (see also § 5).

3. New generation transistor structures.

One of the first devices to benefit from the refinement of the epitaxial growth of GaAs/Al$_x$Ga$_{1-x}$As was the bipolar transistor. Very early on it was realised that significant advantages would accrue by forming the emitter from a wider band-gap material (in this case Al$_x$Ga$_{1-x}$As) than the base, but only recently has the necessary growth technology become available to achieve this in an adequate way. The advantages can be clearly seen from the following expression for the DC current gain (the ratio of the collector current to the hole current flowing from the base to the emitter),

$$\beta \approx \frac{n_e v_e}{p_b v_p} \exp\left(\frac{\Delta E_g}{kT}\right) \qquad (13)$$

where n_e and p_b are the electron and hole densities in the emitter and base respectively, v_e and v_p are the electron and hole effective velocities and ΔE_g is the difference in the band-gap between the emitter and base [1]. This expression is not exact but contains the important physics, namely that the presence of a wider band-gap emitter can significantly enhance the current gain by effectively reducing the base current drawn for a given emitter efficiency (holes are hampered from flowing into the emitter by the enhanced valence-band offset at the heterojunction interface). By grading the alloy over ~ 200 Å near the emitter-base interface one can reduce the conduction-band discontinuity to some extent, so further improving emitter efficiency. The great attraction of the GaAs/Al$_x$Ga$_{1-x}$As materials system is that the degree of lattice matching is almost perfect over the entire alloy composition range and that sizable band-offsets can easily be obtained [9]. For example, with an alloy composition of x ~ 0.2, ΔE_g ~ 250 meV and the exponential factor in Eq. (13) is as large as 1.5 x 10^4 at room temperature.

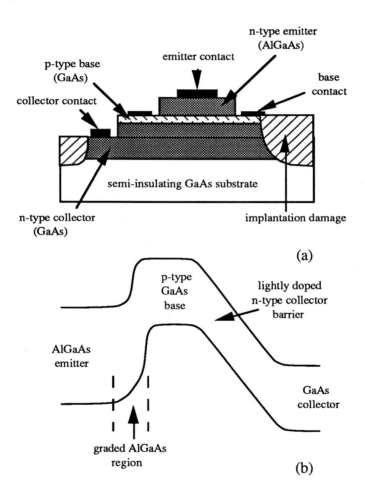

Fig. 11. (a) Schematic of the GaAs/AlGaAs HBT structure, (b) variation of the conduction-band and valence-band profiles throughout the device.

The presence of this large exponential factor in the expression for the current gain of a Heterojunction Bipolar Transistor (HBT) allows the other parameters in Eq. (13) to be adjusted favourably whilst maintaining a practical working current gain, usually about 100. Thus the doping density in the base (p_b) can be significantly increased to values of order 10^{20} cm^{-3}, which allows the base resistance to be reduced, thus improving f_{max}. The high values of the base doping allow the base to be made thinner without catastrophically degrading its resistance or causing punch-through to occur. This reduces the electron emitter-collector transit-time which improves f_T. Velocity overshoot effects can reduce the transit-time even further. One can also reduce the emitter doping (n_e) which lowers the emitter junction capacitance. GaAs/Al$_x$Ga$_{1-x}$As HBTs with values of $f_T \geq 100$ GHz and f_{max} approaching 200 GHz have been realised recently [10,11]. The low propagation delay times (less than 10 ps in ring oscillators based on ECL logic) mean that very high speed digital circuits can be fabricated (up to 32-bit microprocessor circuits have been demonstrated), whereas the low 1/f noise, high transconductance and large current carrying capability are attractive features for a variety of high speed analog applications such as power amplifiers and oscillators [12]. GaAs HBT structures can be reproducibly fabricated on semi-insulating substrates (see Fig. 11) which aids in the formation of low capacitance interconnects. They are especially convenient structures for defining integrated driving circuits in GaAs OEICs, where the current carrying capability is especially important.

At the VLSI level, it is unlikely that Si bipolar circuits will be completely replaced by GaAs HBT circuits. Firstly, the sophistication of Si fabrication technology is much greater. Secondly, light pollution arising from carrier recombination in the direct band-gap base of a GaAs HBT can become a serious problem (affecting neighbouring devices) when the packing density is high. Once the ultimate speeds of operation of conventional Si bipolar technology are reached (20 to 30 GHz) [13], a more natural approach would appear to be to try and realise the same advantages in Si through the incorporation of Si$_{1-x}$Ge$_x$ alloys to form the (narrower gap) base, or SiC to form the (wider gap) emitter. This work has been hampered for years by the relatively poor quality of epitaxially grown Si, mainly due to problems associated with lattice mismatch and of providing adequate doping [14,15]. Recently, however, in the case of Si$_{1-x}$Ge$_x$, significant progress in overcoming these difficulties seems to have been made [16,17]. For narrow bases (≤ 1000 Å) the strain can be accommodated elastically (so-called pseudomorphic growth), and devices with values of $\beta \sim 5,000$ and $f_T \sim 50$ GHz have been reported [18,19,20]. The recent demonstration of devices with a 300 Å Si$_{0.8}$Ge$_{0.2}$ boron-doped base (4 x 10^{19} cm^{-3}) offering room temperature values of $\beta \sim 550$, $f_T \sim 42$ GHz and $f_{max} \sim 40$ GHz are further encouragement [21]. It is unlikely that the ultimate speed offered by GaAs HBTs will be reached with Si HBTs, but for the near future this is less important than being able to retain (with hopefully minimum modification) existing Si fabrication facilities. For OEIC applications one would also like to be able fabricate HBT structures in InP based materials, and this too is a continuing area of research and development.

Along with the bipolar transistor, heterojunction technology has had a significant impact on the field-effect transistor as well [22,23]. In Fig. 12 we compare a standard GaAs MESFET structure with that of a so-called High Electron Mobility Transistor (or HEMT). In the MESFET, the action of the gate electrode is to deplete the conduction channel beneath it and so modify the conduction channel width (and hence its current carrying capability). In the HEMT structure, the gate action is more analogous to that in a MOSFET; the channel electrons are confined to move in a narrow layer at the heterojunction interface and the gate modifies their effective concentration. The original motivation for the HEMT device was the realisation that the low-field mobility of electrons at a GaAs/$Al_xGa_{1-x}As$ heterojunction interface can be extraordinarily large at low temperatures (Fig. 13). In the modulation doped structure [24] the dopants are placed within the $Al_xGa_{1-x}As$ only, and are usually separated from the interface by a thin (~ 50 Å) spacer layer. Electrons from the donors spill-over and are confined in a narrow (< 50 Å) quantum-well at the interface (resulting from the heterojunction conduction-band discontinuity and band-bending due to electrostatic forces), and since they are free to move only in the plane of the interface they are usually referred to as a two-dimensional electron gas (2DEG). The very high mobility found at low temperature is due to the high crystalline quality of the epitaxial layers but more importantly, to the drastically reduced Coulombic scattering from the ionised donors which are now spatially removed from the carriers. In optimum cases, mobilities $\geq 10^7$ $cm^2V^{-1}s^{-1}$ have been reported at ~ 0.35 K [25].

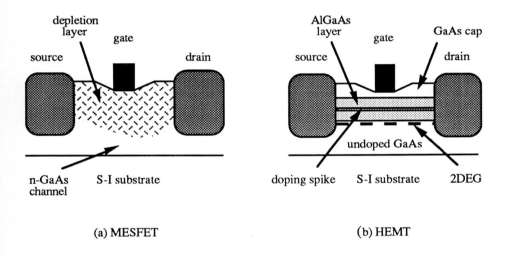

Fig. 12. Comparison of (a) a GaAs MESFET structure, (b) a GaAs/AlGaAs HEMT structure.

The advantages of HEMTs over MESFETs are more subtle, however, than simply assuming that very high low temperature mobility translates into significantly higher speeds of operation. Firstly, the difference in room temperature mobility (where phonon scattering is important) is much smaller (typically only a factor of 2 improvement at best, see Fig. 13) [26]. Secondly, as stressed in § 2.1, usually it is the high-field characteristics that are more relevant than the low-field mobility. A simple analysis is sufficient to confirm the subtle rather than drastic nature of the improvements [27]. If we assume that the maximum possible carrier velocity is the saturated velocity v_s and that the maximum electric field that can be supported is E_b (the dielectric breakdown strength), for a device with gate-length L we have $f_T \sim v_s/2\pi L$ and the maximum source-drain voltage $V_m \sim E_bL$. Similarly, the maximum output power that can be delivered to a load impedance Z is given by $P = V_m^2/8Z$.

One can then relate these important performance parameters to $E_b v_s$ (a product which is material rather than structure dependent) as follows,

$$V_m f_T = \frac{E_b v_s}{2\pi} \tag{14}$$

$$PZf_T^2 = \frac{(E_b v_s)^2}{32\pi^2} \tag{15}$$

basically irrespective of whether the device is a MESFET, a HEMT, or even a bipolar structure. This significantly constrains the values that each parameter can attain in practice without moving to a different materials system.

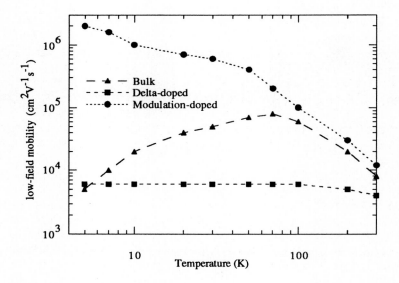

Fig. 13. The low-field mobility for various GaAs structures as a function of ambient temperature (see e.g. [26]).

Nevertheless, the values of f_T and g_m do tend to be somewhat higher in GaAs/Al$_x$Ga$_{1-x}$As HEMT devices with gate lengths in the range ~ 0.1 μm to ~ 1 μm (the typical improvement being a factor of about 40 % [1]) and this is significant. High speed digital HEMT circuits constructed in DCFL logic have demonstrated room temperature ring oscillator delays as low as 10 ps and static Random Access Memory (sRAM) access times below 1 ns. This speed of response is slightly faster than can currently be achieved using MESFET circuits (note that the standard access time for large Si memory chips is ~ 70 ns). The most significant improvement of HEMTs over MESFETs, however, is in the noise figure. Recalling Eq. (3), the explanation for the low value of NF in HEMT devices lies in the higher value of f_T and the lower value of the source-gate resistance R_s (which arises in a low field region where the benefits arising from the higher mobility are more clearly felt). At 10 GHz, the noise figure for GaAs/$Al_xGa_{1-x}As$ HEMTs can be as low as 0.6 dB, which is ~ 0.4 dB less than a state-of-the-art figure for MESFETs [28,29]. At 60 GHz the improvement in NF is typically ≥ 1 dB. This is very attractive for realising high sensitivity microwave front-end receivers, and HEMTs are already widely used in direct broadcast satellite receivers for example.

The performance of discrete HEMT devices can be improved further by utilising different materials systems. Two areas of research are particularly promising. In the GaAs p-HEMT (p stands for pseudomorphic) a thin layer of strained $In_xGa_{1-x}As$ (~ 10 % In concentration and < 100 Å thick) is grown at the GaAs/$Al_xGa_{1-x}As$ interface. Since $In_xGa_{1-x}As$ has a narrower band-gap than GaAs this results in greater carrier confinement in the 2DEG, which can double the value of f_T and also reduce the problems of carriers becoming trapped in the substrate when subjected to large electric fields. The microwave performance of p-HEMTs can readily attain ≤ 0.6 dB noise figure at 12 GHz (with ≥ 11 dB of associated gain). Using InP as a substrate material, HEMT structures in the lattice-matched $Al_{0.48}In_{0.52}As$ / $In_{0.53}Ga_{0.47}As$ materials system can be fabricated which combine the benefits of greater carrier confinement (improving g_m) and a lower effective carrier mass (and hence higher mobility). In devices with gate lengths in the range 0.1 μm to 0.2 μm values of f_T ~ 170 GHz and NF ~ 0.8 dB at 63 GHz (associated gain of 8.7 dB) have been demonstrated [30,31]. Pseudomorphic HEMTs on InP substrates have recently exhibited exceptional performance; i.e. 1.3 dB noise figure and 8.2 dB associated gain at 95 GHz, and as much as 7.3 dB gain at 141.5 GHz [32]. Attempts have also been made to fabricate HEMT structures in the $Si/Si_{1-x}Ge_x$ system, which may point the way to new generation CMOS and BICMOS technologies [33,34]. Here the materials problems referred to in the case of the Si HBT have also proved problematic, but recent results showing electron mobilities of ~ 1.7×10^5 $cm^2V^{-1}s^{-1}$ at 1.5 K are encouraging [35]. To date, the best hole mobilities demonstrated (using pure Ge conduction channels) are ~ 9,000 $cm^2V^{-1}s^{-1}$ at 77K [34].

Epitaxial growth has clearly led to significant improvements in terms of the speed of response of both FET and bipolar devices, coupled with additional improvements, such as very low noise in the case of the HEMT. If one is prepared to operate the devices at lower temperatures (such as 77 K, the boiling point of liquid nitrogen), then even more significant advantages can be gained. For instance, at 12.5 K the noise figure for a HEMT at 8.5 GHz can be less than 0.1 dB [28], reflecting the greatly increased mobility at such temperatures. This is much lower than even cryogenically cooled GaAs MESFETs can attain, which for years have provided the basis for ultra-high sensitivity microwave receivers. The transit-time frequency also increases accordingly. Whereas for specialised high speed applications in the past it has proved possible (and profitable) to cool Si CMOS and GaAs MESFET circuits, this has not been possible with Si bipolar circuits because of the poor low temperature characteristics of these devices (related, amongst other things, to minority carrier freeze-out). HBT structures, which allow for much higher base doping levels, circumvent this problem to a large extent, and it is interesting to note that the measured ECL gate delays of Si HBTs at 84 K (~ 28 ps) are practically identical to those at room temperature [36]. We shall return to the potential benefits of low temperature electronics in a later section (§ 6).

For many years attempts have been made to realise a fast, vertical *unipolar* transistor. Such a transistor could (in principle) offer a number of advantages compared to bipolar transistors, such as removing problems associated with minority carrier recombination, reducing the switching voltages and powers for logic operations (since only voltages corresponding to heterojunction offsets rather than band-gaps are typically required) and eliminating a troublesome surface recombination term which plagues attempts to scale down bipolar structures (this affects the hole velocity in Eq. (13)). Since these structures tend to utilise carriers significantly disturbed from their equilibrium carrier distributions they tend to be referred to collectively as Hot Electron Transistors (HETs). Many epitaxially grown GaAs/Al$_x$Ga$_{1-x}$As versions of the basic HET idea have been studied since the mid 1980's [37,38,39,40], and the conduction-band profile of one such structure is depicted in Fig. 14.

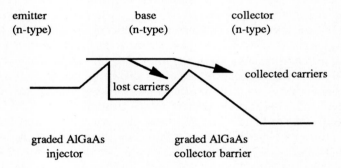

Fig. 14. Schematic of the conduction-band profile of GaAs/AlGaAs Hot Electron Transistor (HET). Carriers 'launched' into the base region have an excess energy corresponding to the emitter-base heterojunction offset; in crossing the base some lose sufficient energy to be unable to surmount the collector barrier.

To date, however, the performance of these devices has been disappointing. The time taken to charge the base-emitter capacitance can be written as follows [1],

$$\tau_b \sim \frac{\varepsilon L^2}{L_e n_{2D} \mu} \quad (16)$$

where L is a typical base width, L_e is the emitter barrier thickness (~ 0.1 μm), μ is the electron mobility and n_{2D} is the charge density per unit area in the base. For high speed of operation and high collector efficiency α (the fraction of carriers launched across the base that reach the collector, which determines the current gain $\beta \sim \alpha/(1-\alpha)$) one wants a narrow base and therefore high doping level to compensate (for the increased base resistance). Unfortunately, for n-type doping levels in excess of $\sim 10^{18}$ cm^{-3} (in GaAs) electron scattering in the base becomes significant. Even worse, hot carriers injected into such a heavily doped base rapidly lose their excess energy to the emission of coupled optic phonon-plasmon modes with energies ~ 60 meV [41]. If the 'launching' energy is 0.3 eV (one cannot go higher with GaAs or else intervalley transfer significantly reduces the carrier transit-time [42]), the typical carrier mean free path is only ~ 20 nm. For acceptable collector efficiencies (and hence current gains) the base must therefore be less than about 40 nm wide, and it is effectively impossible to reconcile the constraints implied by the above for high speed operation. The largest room temperature current gains reported for GaAs HETs are only about 9 or 10. Better results for HETs fabricated in different material systems have been obtained (see § 6), and also for modified GaAs structures whereby the base electrons are forced into a 2DEG at the base-collector interface [1,43]. The plasmon-mode energies and dispersion relations of such a layer of electrons are quite different to those of a (3D) bulk-doped base which is an advantage. The disadvantage is the naturally high resistance of the narrow base and the exceptional technological problems inherent in forming a selective ohmic base contact. No HET structure to date has demonstrated a level of performance which is suitable for commercial exploitation.

We close this section on transistor structures by briefly mentioning two other avenues of research which are, as yet, unproven. MBE growth allows very narrow, very highly doped regions of semiconductor to be accurately defined; referred to as planar doping or 'spike' doping or, in the limit where the dopants are deposited in practically a single atomic plane, delta-doping [44]. The delta-doped FET (δ-FET) uses a delta-doped layer to form a high carrier density conduction channel which can then be modulated as in a MESFET, and in principle the high carrier densities should lead to high values of g_m (the typical mobility of a GaAs delta-layer is shown in Fig. 13) [45,46,47]. The concept of real space transfer (RST), whereby carriers travelling in a high mobility heterojunction channel (e.g. GaAs) are heated by the electric field and transferred into the lower mobility barrier material (e.g. $Al_xGa_{1-x}As$), leading to NDR [48], has been exploited to make transistors like the CHINT (Charge Injection Transistor) and the NERFET (Negative Resistance Field Effect Transistor) [1]. With these transistors, values of $f_T \sim 30$ GHz ($f_{max} \sim 10$ GHz) and $f_T \sim 60$ GHz ($f_{max} \sim 18$ GHz) have been demonstrated using the GaAs/AlGaAs and pseudomorphic InGaAs / AlGaAs / GaAs materials systems respectively [49,50]. The most attractive potential use of RST transistors is in utilising the non-monotonic nature of the transistor characteristics to derive novel logic operations [51] (see also § 6).

4. New generation microwave diodes.

The last few years have seen the development of a new class of diode to rival the traditional Schottky diode for microwave mixer and detector applications. The planar doped barrier (or PDB) diode [52] (and the closely related Camel diode [53]) are bulk, unipolar structures, whereby the barrier governing thermionic emission arises not from the band-discontinuity arising at a metal-semiconductor interface as in the Schottky diode, but from the full depletion of a thin (typically ≤ 100 Å) p-type layer (in the case of majority carrier electron devices) placed within the intrinsic (i) region of an n - i - n doping structure (see Fig. 15). To obtain a rectifying I-V characteristic, the p-type doping spike is placed close to one end of the intrinsic region ($L_2 \ll L_1$, with L_1 typically ≤ 0.4 μm). The difficulty of realising such a device using conventional ion implantation meant that this device concept could only be thoroughly explored once the epitaxial growth of GaAs was 'perfected'.

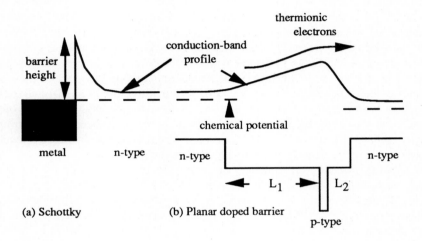

Fig. 15. A schematic of the planar doped barrier diode conduction-band profile (under small forward bias) compared to that of a Schottky diode (unbiased). The fully depleted p-type layer in the structure on the right is negatively charged, and so represents a potential barrier to the majority carrier electrons.

The original motivation for pursuing the PDB diode as a microwave element lay in its great flexibility. For instance, in a very asymmetric structure ($L_2 \ll L_1$) the barrier height within a depletion model may be written as follows,

$$\varphi_0 \approx \frac{e^2 N_p t}{\varepsilon}\left\{L_2 + \frac{t}{2}\left(1 + \frac{N_p}{N_d}\right)\right\} - \mu \qquad (17)$$

where N_a and N_d are the acceptor and donor concentrations in the p-type and n-type layers respectively, t is the thickness of the p-layer and μ is the position of the chemical potential (relative to the bottom of the conduction-band). It is very easy to adjust the barrier height (and hence the value of the forward 'turn-on' voltage) simply by controlling the thickness and doping of the p-type layer during growth. In particular, it is easy to realise devices with a low barrier height (≤ 300 meV) which are useful for zero-bias detector and low-drive mixer applications. By contrast, fabricating Schottky diodes with comparable barrier heights is usually very difficult to do reproducibly (through partial annealing for example), since the barrier height of metals such as Au and Al on GaAs tend to 'pinned' to values ~ 0.9 eV by the presence of a large density of surface states in GaAs. Another advantage lies in the fact that the capacitance of a PDB diode is practically independent of the applied bias [52], which contrasts strongly with the Schottky diode case and makes for easier circuit matching.

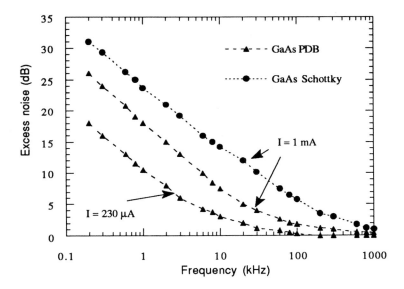

Fig. 16. Comparison of the excess noise of GaAs PDB and Schottky diodes with the drive current levels under which the measurements were taken indicated (after [54]).

Recent research has seen a very detailed characterisation of the performance of carefully designed GaAs PDB diodes in microwave applications up to 100 GHz [54]. As far as detector applications are concerned, superior tangential sensitivity and voltage sensitivity (output voltage for a given input power) compared to state-of-the-art Schottky diodes has been demonstrated at all frequencies. At 35 GHz for example, tangential sensitivities of ~ - 53 dBm and voltage sensitivities (at - 20 dBm input power) of ≥ 2,700 mV/mW are readily achievable. The temperature stability (change in performance with temperature) of PDB diodes is also superior; typically only ~ 1 dB variation over the range - 40 ^{0}C to + 80^{0}C compared to ~ 3 dB for the Schottky diode. This is at first sight surprising, since both devices exploit thermionic emission which is a very temperature dependent process. Thus the current is usually written in the form,

$$I = I_s(V,T)\left\{\exp\left(\frac{eV}{kT}\right) - 1\right\} \tag{18}$$

where the reverse-bias saturation current I_s (which is bias dependent to some extent) depends exponentially upon the barrier height to temperature ratio. The improved temperature stability results from a compensating effect in the PDB diode whereby there is actually a slight *increase* in the barrier height with temperature, thus limiting the variation of I_s to some extent. Such an effect in Schottky diodes is absent because of Fermi-level pinning due to surface states at the interface. For mixer applications, the low conversion loss (≤ 4.5 dB at 10 GHz) at low local oscillator (LO) power levels (less than 0.5 mW) is attractive for high frequency applications where very often providing sufficient LO power is problematic. The absence of defects and traps associated with the metal-semiconductor interface also leads to very substantial improvements in terms of resilience to incident pulsed-power (~ 10^3 times the burnout capability of unprotected Schottky diodes) and lowering of 1/f noise (recall the arguments in § 2.2). The truly exceptional 1/f noise performance is illustrated in Fig. 16, with a corner frequency as low as 10 kHz, which is very important for applications such as doppler-shift radar for example.

The ultimate speed of response of PDB diodes is still not fully understood. The one drawback of the PDB diode is that its series resistance R_s tends to be quite high, resulting from the naturally resistive intrinsic regions and hot electron injection effects as electrons surmount the abrupt barrier at the p-type layer. Estimates based on Monte Carlo simulations [55] suggest that a realistic mixer cut-off frequency (§ 2.2) of up to 300 GHz may be attainable, but probably not much higher. For any future applications in the 500 GHz to 1 THz regime the Schottky diode will probably remain the first choice component, although one expects to see an increasing use of PDB diodes below 100 GHz.

A fundamental limitation of transit-time diodes for microwave power generation is that as the frequency goes up, so the devices get shorter and less powerful and efficient as a result. The heterojunction Gunn diode (depicted in Fig. 17) ensures that the maximum available transit-length is fully exploited by arranging for the carriers to be injected into the drift-region with just about the right amount of energy (~ 0.3 eV) to instigate intervalley transfer straight away [56,57]. In a conventional structure, the electrons have to be heated by the applied field to gain this energy which results in a dissipative region (or 'dead-zone') at the start of the drift-region. This limits efficiency (power is *dissipated* here) and in short, high frequency devices, the dead-zone takes up an unacceptable fraction of the total transit length. The performance of the heterojunction device is markedly superior [58], with power levels (continuous wave, or CW) of ~ 80 mW at 94 GHz and conversion efficiencies of ≥ 4 %. The temperature stability is also improved and the threshold voltages are reduced, as now the intervalley transfer process has been made easier to 'accomplish' by exploiting the built-in electric field due to the heterojunction. The purpose of the n-type doping spike adjacent to the graded $Al_xGa_{1-x}As$ injector is to tailor the electric field so that it has the correct value (~ 5 to 10 kV cm^{-1}) in the drift-region to ensure stable domain propagation. It also serves a secondary purpose by acting as a well defined nucleation site for domain formation, ensuring that each domain traverses basically the same length of drift region. This means that the frequency spread of the output power (or FM noise) in these devices is usually found to be very low (e.g. ≤ - 88 dB$_c$/Hz at 100 kHz off-carrier, where the subscript c means with respect to the carrier). The heterojunction Gunn diode is the first, commercially available device which exploits hot electron injection in a truly fundamental (as opposed to peripheral) way.

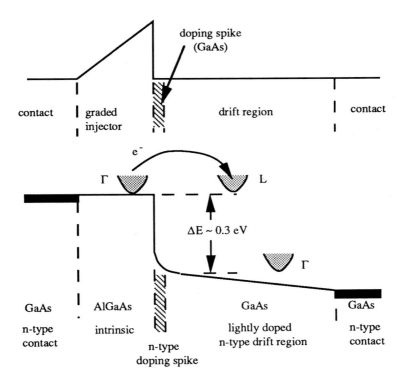

Fig. 17. The heterojunction Gunn diode concept showing how carriers launched into the drift-region acquire sufficient energy to instigate intervalley transfer.

The ability to define doping profiles accurately using MBE and MOCVD has enabled high performance IMPATT diodes for frequencies of operation above 30 GHz to be reliably manufactured. One Watt of (CW) output power at 94 GHz is now possible with conversion efficiencies of between 5 and 10 %, and Si double-drift structures have recently been shown to be capable of 42 W peak pulsed-power operation at 96 GHz [59]. For high power radar applications the IMPATT diode is likely to remain the most important solid-state power source, but for communications purposes the noise levels are usually unacceptable and an alternative is required. The most important candidate is the heterojunction Gunn diode mentioned above, which should be capable of producing meaningful quantities of power up to frequencies of ~

200 GHz (for GaAs devices). Other options are being studied, however. In particular, there are requirements for sources which combine low power (but high efficiency) with low noise. The loss of power is not necessarily catastrophic, as long as a reasonable signal-to-noise ratio can be achieved and one has recourse to high sensitivity receivers of the type discussed already. The high efficiency is often an important criterion in applications such as satellite communication links where power is at a premium.

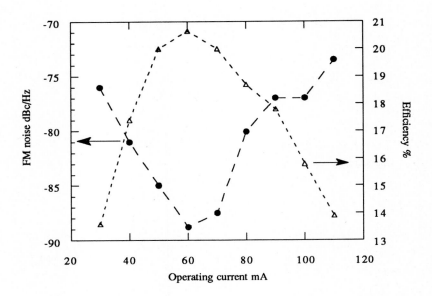

Fig. 18. FM noise and conversion efficiency as a function of drive current for a GaAs Hi-Lo single-drift IMPATT diode. The output power at 60 mA is ~ 300 mW, and the frequency of operation is ~ 30 GHz (after [60]).

By carefully designing the doping profile of an IMPATT diode structure one can often improve the efficiency and noise at the expense of reducing the potential output power. Fig. 18 shows the efficiency and FM noise of a purpose designed 30 GHz GaAs single-drift IMPATT diode which generates in excess of 300 mW at the maximum efficiency drive current (60 mA), which also happens to be the minimum noise point [60]. The levels of efficiency, noise and power are attractive for these frequencies, although the AM noise (which is a traditional problem with IMPATT devices and has not been measured here) may still be poor. Gunn diodes, which are strongly non-linear, generally have much superior AM noise performance. More radical restructuring of the avalanche transit-time diode has been suggested through the incorporation of heterojunction elements, but with no significant success to date, and no fundamental guiding principle to suggest that substantial progress in this direction will be forthcoming in the immediate future. If one is prepared to sacrifice most of the available power, however, then progress can be made. In devices whose doping levels are such that a significant fraction (or even all) of the current at breakdown is due to interband *tunnelling*, as opposed to avalanching, then the noise associated with the statistical nature and rapid gain of the avalanche process itself can be substantially reduced. The drawback is that the current-voltage phase relation in MITATT (Mixed Tunnel and Avalanche Transit-Time) and TUNNETT (Tunnel Transit-Time) diodes is less favourable for efficient power generation [61,62]. Performance demonstrated to date includes 25 mW output power (2.8 % efficiency) at 70 GHz [63] and 32 mW (2.6 % efficiency) at 93.5 GHz [64] with GaAs TUNNETT devices, and 3 mW (0.5 % efficiency) at 150 GHz with a GaAs MITATT diode [65]. Although no noise measurements were directly reported for these devices, significantly lower noise has been observed in $Si/Si_{1-x}Ge_x$ MITATT diodes generating ~ 25 mW at 103 GHz [66]. In principle, it should be possible to extend the ultimate frequency of operation of these devices up to about 300 GHz, although here the issues of cavity design become pressing (see also § 6).

5. New generation optoelectronic devices.

Epitaxial growth has led to two important developments in the field of semiconductor laser technology. The quantum-well (QW) laser [67] is similar in construction to the double heterostructure (DH) laser discussed in § 2.3, but with a quantum-well (or a number of quantum-wells) grown within the active region of the device. As in the case of the relative merits of the HEMT over the MESFET, the advantages gained from this are rather subtle. Firstly, the wavelength of operation can be tuned to some extent by altering the well widths (the dominant recombination occurs between the lowest electron and heavy-hole sub-bands, whose relative energy separation is well-width dependent). More importantly, the two-dimensional nature of the density of electron and hole states in the wells results in an improved variation of gain with drive current, a further reduction in threshold current and a reduced temperature dependence of the threshold current. It is usual to write,

$$J_{th} = J_0 \exp(T/T_0) \qquad (19)$$

where a large value of T_0 implies a more temperature stable performance. For a $GaAs/Al_xGa_{1-x}As$ DH structure operating at ~ 0.8 μm, T_0 might be of order 150 K, whereas for a QW laser structure emitting at the same wavelength T_0 might perhaps be as high as 220 K [67]. Values of $J_{th} \leq 50$ A cm^{-2} have been reported for QW lasers [68]. The reason for retaining the DH 'cladding' structure around the quantum wells (see Fig. 19) is to maintain the optical wave confinement and waveguiding properties.

In conventional semiconductor laser diodes the light is emitted in a direction perpendicular to that of the current flow. In recent years [69,70] it has proved possible to grow multilayer structures which act as mirrors, enabling the cavity to be formed in the *vertical* rather than transverse direction (the Vertical Cavity Surface Emitting Laser or VCSEL). The mirror regions consist of alternate layers of GaAs and $Al_xGa_{1-x}As$

(for example), each of which is about $\lambda/4n \sim 700$ Å thick (where n is the refractive index). This, coupled with the slight dielectric mismatch between GaAs and $Al_xGa_{1-x}As$, is sufficient to form a quarter-wave stack mirror, and mirrors with between 10 and 20 layers can be made ≥ 99.8 % reflecting. The active region between the mirrors is typically $\sim \lambda/n \sim 3,000$ Å thick, a much narrower cavity length than in a conventional transverse laser structure (~ 450 μm), which means that the VCSEL tends to have fewer cavity modes, which can lead to a reduced spectral linewidth. The great attraction of VCSEL technology for the future is that once the material is grown, it is then relatively easy to fabricate very small laser diodes or to provide for very high monolithic integration densities ($\geq 10^6$ cm^{-2}), leading to high overall output powers. One can also envisage a whole new range of OEICs benefiting from the ease of integration offered by the VCSEL structure. The drawback lies in the inherent difficulty of growing the structure and in the fact that the mirror regions also tend to be electrically resistive, which increases the voltage levels and threshold currents (and hence the power dissipation). Nevertheless, by suitable grading or heavily doping the mirror regions, values of $J_{th} \leq 1$ kA cm^{-2} can probably be realised.

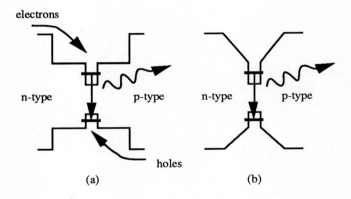

Fig. 19. Two examples of a quantum-well laser band-profile, (a) featuring abrupt composition changes, (b) featuring continuous grading.

The modulation bandwidth of a semiconductor laser depends upon its differential gain, a measure of how quickly the light intensity responds to small changes in carrier density (i.e. current). The differential gain of a QW laser near threshold is greater than that of a conventional DH laser because of the modified density of states, and this favours higher speed operation (as long as the output power levels can be maintained). It may eventually prove possible to reach bandwidths approaching 50 GHz in fully optimised structures [1].

Heterojunction technology has played an important role in improving the performance of photodetectors as well as lasers. $GaAs/Al_xGa_{1-x}As$ PIN diodes are more efficient than Si devices in the range 0.65 μm to 0.85 μm, where the direct band-gap of GaAs is favourably exploited to enhance absorption and the $Al_xGa_{1-x}As$ layer (with its wider band-gap) forms a transparent contact region which allows light to reach the absorption region unattenuated [71]. The lattice matched $In_{0.53}Ga_{0.47}As/InP$ combination is ideal for longer wavelength detection (here the InP substrates are transparent), and at 1.3 μm modulation bandwidths of almost 20 GHz have been demonstrated combined with efficiencies of ≥ 80 % [72]. One of the most impressive examples of improved performance, however, occurs in the case of the so-called Separate Absorption and Multiplication (SAM) avalanche photodiode (APD) depicted in Fig. 20.

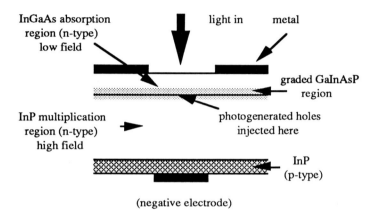

Fig. 20. Schematic of a separate absorption and multiplication (SAM) avalanche photodiode for longer wavelength detection (1.3 μm to 1.6 μm).

In this structure, the light is absorbed in the $In_{0.53}Ga_{0.47}As$ layer which is a low field region; this reduces the amount of interband tunnelling and so keeps the dark current down. The carriers (holes in this case) then enter the InP region, which is under a high electric field, where they rapidly avalanche (the large band-gap of InP helps to reduce tunnelling in this high field region). The point of using holes as the injecting species to initiate avalanche multiplication is that in InP the hole ionisation rate is about three times that of the electrons, and so this arrangement can be exploited to reduce the excess noise factor F(M) (see § 2.3). To gain maximum performance the doping profile must be very carefully optimised so that the electric field in the $In_{0.53}Ga_{0.47}As$ layer is not too large, but not too small either (otherwise slow hole diffusion times limit the speed of response). It is also necessary to grade the heterojunction interface using the $Ga_xIn_{1-x}As_yP_{1-y}$ quaternary alloy to prevent holes becoming trapped [73,74]. Nevertheless, SAM APDs have been made which exhibit low dark currents, improved signal-to-noise ratios and gain-bandwidth products of ~ 30 GHz [75]. The performance of these devices in the 1.3 μm to 1.6 μm wavelength range is superior to the Ge APDs they are designed to replace.

Initial experimental observations of the avalanche current resulting in GaAs/$Al_xGa_{1-x}As$ multiple quantum-well structures when initiated first by holes and then electrons appeared to suggest a significant asymmetry in the ionisation rates of the two types of carrier, apparently resulting from the different heterojunction discontinuities with the conduction and valence bands [76]. This created considerable excitement at the time in terms of possibly being able to 'engineer' an entirely new class of low noise APD. Unfortunately, today it appears that the offset differences in the GaAs/$Al_xGa_{1-x}As$ materials system are too similar for the effect to be of much significance; a point later confirmed by direct measurements of the noise multiplication factors themselves [77]. In the $In_{0.53}Ga_{0.47}As$ / $In_{0.52}Al_{0.48}As$ / InP system, however, this effect is more pronounced and values of F ~ 2.7 for M ~ 10 have been measured, with a bandwidth of 9.3 GHz [78].

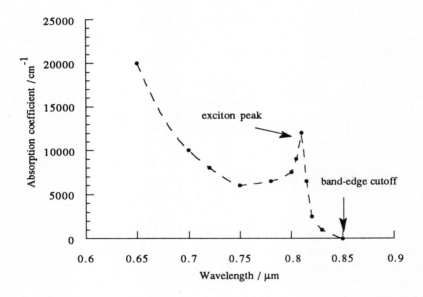

Fig. 21. Schematic graph showing the gross features of the room temperature absorption spectrum for a GaAs/AlGaAs multiple quantum-well (MQW) structure (see [79] for more detailed results).

Whereas the dominant driving force behind optoelectronics has traditionally been telecommunications, this is slowly being supplemented by requirements in the world of high speed computing. We have already referred to the prospect of using optical interconnects to provide high speed data transfer. Whilst not necessarily used for their intrinsic speed, semiconductor lasers are becoming ever more important in printing and data storage (CD-ROMs). As optics becomes more widespread, so the attraction of integrated optoelectronics grows, helped by the development of sources like the VCSEL. The absorption properties of multiple quantum-well (MQW) structures are also attractive in this respect [79], since they are greatly affected by the carrier confinement in the wells. Recent developments, for example, have seen GaAs/ $Al_xGa_{1-x}As$ MQW infrared detectors developed for the 8 μm to 12 μm regime based on *inter sub-band* transitions of electrons within the conduction band, offering good single pixel performance and excellent uniformity for forming large 2D staring arrays, posing a serious alternative to the narrow band-gap Cadmium Mercury Telluride (CMT) devices currently employed [80,81]. One of the more interesting observations is the presence of a strong exciton absorption peak (even at room temperatures) in MQW structures for wavelengths near to that corresponding to the band-gap energy (see Fig. 21). This strong peak is due to the spatial confinement of the electrons and holes within the narrow wells, which substantially increases the exciton binding energy over that of the bulk material thus hindering exciton dissociation. From an applications point of view, two associated features of this exciton absorption peak are of possible interest. Firstly, it can be saturated by applying an incident laser intensity ~ 500 W cm^{-2} (in the case of GaAs/Al$_x$Ga$_{1-x}$As), leading to a reduction in the absorption coefficient α of perhaps 5,000 cm^{-1} [82]. Although the implied non-linearity is significantly greater than that found in the bulk, few direct applications have been found. Secondly, because the shape of the QW confining potential is changed by applying an electric field perpendicular to the growth direction, the electron and hole sub-band energies (and hence the position of the exciton absorption peak) can be shifted in wavelength by perhaps 10 nm [83]. This *electroabsorption* effect is also much stronger than in the bulk because the QW confines the carriers and inhibits exciton dissociation through tunnelling, permitting larger fields to be applied. Since moving the exciton peak changes the absorption for wavelengths near the peak, it can be made the basis of a QW modulator [84,85].

An optical modulator is a medium whose transmission properties can be altered (in this case electrically), thus enabling the intensity of, say, a laser, to be indirectly modulated. The potential advantages of this approach as compared to directly modulating the laser drive current include removing the difficulty of controlling high speed current swings and removing the output chirp which results from the fact that the laser wavelength is actually current dependent to some extent. A MQW modulator consists of a sequence of up to perhaps 50 quantum-wells sandwiched between n and p-type semiconductor layers which permit large reverse bias fields to be applied (see Fig. 22). By changing the applied bias one can move the exciton peak around and so change the transmission; since the earliest demonstration, contrast ratios exceeding 10:1 (the difference in intensity between the high (on) and low (off) transmission states) and switching (modulation) speeds below 100 ps have been reported [86]. There is still some chirp associated with a small change in refractive index with bias, but this is typically reduced by a factor ~ 3 over that incurred through direct modulation [87]. The big drawback of MQW modulators is that they are lossy and very wavelength specific (the laser radiation wavelength must almost coincide with that of the exciton peak since the effective tuning range is not large). Conventional electrooptic materials such as Lithium Niobate (LiNbO$_3$), which exploit a change in refractive index with bias rather than absorption, are much less lossy and also much less wavelength specific. On the other hand, devices made from LiNbO$_3$ are usually much more bulky to achieve the same contrast ratio, and are clearly less easily integrated with on-chip electronics than MQW structures.

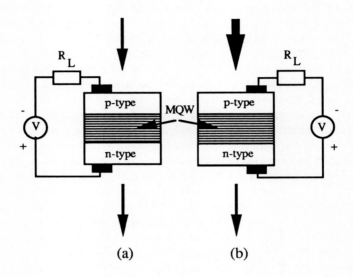

Fig. 22. A MQW modulator configured as a SEED showing (a) low current, high transmission state, (b) high current, low transmission state.

Since the MQW modulator structure is basically a reverse biased PIN diode, it is also capable of acting as a photodetector (§ 2.3). Under certain circumstances one can exploit the twin properties of photodetection and bias dependent absorption to form an optically bistable element called a Self-Electro-Optic-Device (or SEED) [88]. This works as follows. For low light levels the photocurrent will be low, and in a circuit consisting of a power source, MQW modulator and resistive load (see Fig. 22), most of the voltage will be dropped across the modulator, which is designed to be in a high transmission state under large bias. As the light intensity changes, so does the photocurrent and the voltage across the load, causing the modulator voltage to change and the absorption to change. As long as the responsivity of the modulator is sufficiently non-monotonic then the device can switch between a stable high transmission state and a stable low transmission state, depending upon the incident intensity. In more sophisticated circuits one can therefore use one light beam to control another, and the SEED behaves as an optical transistor performing the NOR function. Switching rates ≥ 1 GHz have been demonstrated with switching energies ≤ 20 fJ/μm^2 (the switching speed scales inversely with the switching power) [89]. One of the first applications where arrays of SEED devices (spatial light modulators) have been actively explored is as part of an all optical switch and interconnect technology for multiple fibre networks. These SEED arrays are also being studied for use in ultra high speed (and highly parallel) all-optical processing architectures. Although perhaps it is still too early to say with any certainty, these devices may well be amongst the first radically new 'quantum' devices (in the sense that they exploit small-scale physics to perform a function quite distinct from that in any 'conventional' semiconductor device) to emerge which offer a *practical* level of performance.

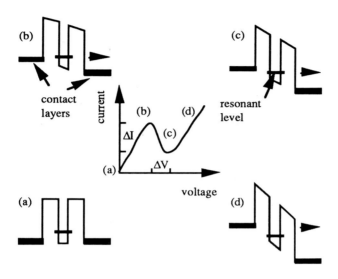

Fig. 23. Showing the conduction-band profile of the double-barrier resonant tunnel diode under various degrees of applied bias and the resulting form of the current-voltage characteristics, which exhibit NDR.

6. Resonant tunnelling.

Of all the novel quantum phenomena realisable in carefully grown semiconductor multilayer structures, that of resonant tunnelling [90,91] has probably received the greatest attention, and we end these lectures with a brief discussion of the implications it holds for high speed device operation. The concept of quantum-mechanical tunnelling is well known of course, and in the double barrier structure depicted in Fig. 23 electrons within the conduction-band can tunnel through both barriers allowing a current to flow even when the classical 'resistance' would appear to be infinite. Within the standard WKB approximation the tunnelling probability for an electron to pass through a single barrier depends exponentially on the barrier width and the incident electron energy, being very small for thick barriers and low energies. Naively one might expect that the probability to tunnel through both barriers would be the product of two such exponential factors. However, when the distance between the barriers is small (a well width ≤ 100 Å say), one needs to consider the effects of multiple reflections within the well. The surprising fact is that for incident energies very close to what would be the energy of the quantum-well bound states if the barriers were very thick, the transmission coefficient or tunnelling probability can approach unity (resonant tunnelling) [92].

The device implications follow from considering the fact that as the bias is increased and the chemical potential in the contact layer is brought into and then out of resonance with the quasi-bound levels, so the current increases, then falls, and then increases again (see Fig. 23). As might be expected, the observed NDR is very sensitive to temperature and is much more pronounced at low temperatures where thermal broadening and scattering are reduced. Nevertheless, current swing peak-to-valley ratios as high as 4:1 in GaAs/Al$_x$Ga$_{1-x}$As [93] and 7:1 in the InP lattice-matched materials [94] have been observed at 300 K. With respect to Fig. 23, the maximum power that one might expect to obtain from a Double Barrier Resonant Tunnel (DBRT) diode oscillator is given by [95],

$$P_{max} \approx \frac{3}{16} \Delta I \Delta V \tag{20}$$

Assuming that current densities as high as 10^5 A cm^{-2} can be realised, it turns out that because of the low voltage swings (typically much less than 1 V), the maximum power for a practical device area (see below) is still only a few mW, even at low frequencies, although the conversion efficiencies can be quite high. Nevertheless, as we have stressed before, there are specialist applications where low power and high efficiency are acceptable. One can improve things slightly by combining a drift region with the DBRT structure to make a Quantum-Well Injection Transit-Time (QWITT) diode [96,97]. As a measure of their potential, QWITT diodes fabricated on InP substrates with a 47 Å In$_{0.53}$Ga$_{0.47}$As well, 25 Å strained AlAs barrier layers and ~ 1,800 Å In$_{0.53}$Ga$_{0.47}$As drift region have been reported recently as giving ~ 10 mW output power at ~ 2 GHz with a conversion efficiency approaching 50 % (the highest of any transit-time device operating under continuous conditions) [98].

The cut-off frequency for a tunnel diode is usually defined as the frequency above which the diode no longer exhibits NDR. Based on a simple equivalent circuit model (e.g. Fig. 7(b)) one can write f_c as follows,

$$f_c = \frac{1}{2\pi |R| C} \sqrt{\frac{|R|}{R_s} - 1} \tag{21}$$

where |R| is the modulus of the negative resistance, R_s is the series resistance and C is the junction capacitance. The negative resistance scales inversely with diode area, whilst the capacitance scales with area and inversely with the length of the relevant depletion regions. For high frequency operation a small area is required. Whereas the very high doping levels (and hence short depletion lengths) required in a conventional p-n junction inter-band tunnel diode limit f_c to a maximum of about 100 GHz [1], for DBRT diodes the doping and the capacitance can be made much smaller, suggesting that cut-off frequencies in excess of 1 THz might be attainable. More sophisticated equivalent circuits have been proposed which allow for the time taken for the carriers to tunnel [99,100]. The issue of the carrier transit-time through a tunnel structure remains the subject of much confusion and debate, but it is generally agreed that circuit limitations (rather than transit-time effects) will define the eventual upper frequency of device operation, as least as far as oscillators are concerned [100].

In one of the first demonstrations of high speed operation, mixing and detection were performed (at 25 K) at frequencies of up to 2.5 THz (using infrared lasers), which confirms the existence of non-linearity at these frequencies [101]. To make an oscillator the diode must be combined with a suitable resonator circuit (§ 2.2). The first reported results for oscillator action (the diode being mounted in a co-axial cavity) were an output power of 5 µW (at 200 K) at 18 GHz with a conversion efficiency of 2.4 % [95]. Later work saw the frequency of stable oscillation increased first to 56 GHz (60 µW) and then to 412 GHz (0.2 µW), both at room temperature [102,103]. At these frequencies quite sophisticated waveguide cavities must be used to gain maximum performance. The present record for stable high speed oscillation appears to be 712 GHz, using a DBRT diode fabricated from InAs/AlSb [104]. Although these frequencies are highly impressive the output power levels are still too small to be of practical benefit. Probably the best hope of short-term success is a DBRT technology aimed at providing low power but high efficiency and low noise sources (integrated with other passive microwave components on semi-insulating GaAs substrates) for the frequency range 100 to 200 GHz [105]. These may well prove of use in high performance satellite communication links for example.

The non-monotonic nature of the I-V characteristics of DBRT diodes can be exploited to create new types of multi-state logic and so-called multi-functional devices which are capable of performing several functions within one device. Multi-functionality is attractive because it promises reductions in circuit complexity, which saves space and also allows for an increase in overall circuit speed. On their own, DBRT diodes have been shown to be capable of logic operations such as parity generation; 11-bit parity generation has been demonstrated in a single device consisting of 5 DBRT diodes integrated vertically, and it would take 10 conventional Exclusive-OR gates to carry out the same task [106]. DBRT diodes can also be integrated into transistor structures such as a FET, a bipolar transistor and even a HET (forming the RHET or Resonant tunnelling Hot Electron Transistor, where the resonant tunnelling element forms the emitter barrier, c.f. Fig. 14). Resonant tunnelling bipolar transistors fabricated in lattice-matched InP-based materials have demonstrated values of $f_T \sim 24$ GHz and DC current-gains of ~ 18 dB at 300 K, and a single such device is capable of 4-bit parity generation replacing 24 conventional transistors [107]. Very simple, high speed analog-to-digital converters can be made in principle using these devices. RHETs have been fabricated with (at 77 K) values of f_T ~ 121 GHz and current gains ~ 35 dB at 1 GHz [108]. The RHET can be made to function as a flip-flop circuit which would normally require 6 transistors, amongst other things.

Although these results are encouraging, there are some drawbacks. Firstly, the performance of resonant tunnelling logic devices is much better at 77 K than at room temperature, a reflection on the degradation of the tunnelling process with increasing temperature. For applications where cryogenic cooling is a possibility they start to look more attractive. Secondly, the tolerances required on device fabrication are formidable if the bit-error rate in multi-state logic systems is to be kept small (in many ways 3-state logic is closer to be analog than it is to being binary). Finally, although one can achieve a reduction in circuit complexity, this is bought at the expense of greatly increased cost and inconvenient system redesign. The incorporation of resonant tunnelling elements within FET and bipolar transistors also slows down their speed of operation, but this may be compensated for by the improved circuit speed resulting from reduced complexity.

7. Conclusions.

Conventional semiconductor devices are now reaching the level of sophistication where some new feature is required if the rate of improvement in performance set over the last two decades is to continue. Epitaxial growth technologies such as MBE have allowed substantial improvements to be made to FETs, bipolar transistors and lasers without necessitating wholesale changes in fabrication technology and system configuration, which is a great attraction. At the same time, a number of new devices are beginning to emerge from this approach which have no direct conventional counterpart, although often the functions they perform are standard. Even here, however, there are examples of new devices (some of the resonant tunnelling structures discussed in § 6 for instance) which promise new functions leading to an overall reduction in circuit complexity, and so the hope of an alternative route to higher speed performance. The necessary redesign of system architectures to exploit such devices makes their eventual fate harder to evaluate. At a more speculative level still, a great deal of research has been devoted (see e.g. [109]) and is still being devoted to the device implications of much more 'novel' physics resulting from quantum confinement in not just one, but two and even three dimensions simultaneously (through a combination of epitaxial growth combined with sophisticated lithography). Whether some of this work will prove to be of practical long-term value in a device context is still unclear and a matter for debate. The signs are, however, that there is still some way to go before the full possibilities of semiconductor device technology have been fully explored.

In these lectures we have discussed a number of areas of application where speed of response is an important issue. The aim has been to give the reader a feel for the important physical principles behind high speed device operation, and to show that one should not think of device speed in isolation; factors such as noise, sensitivity, efficiency and power consumption must always be considered in parallel. Perhaps more importantly, one should never think of a device in complete isolation from its intended application, as how it is exploited in a system context often places significant constraints on its performance. At a more mundane level, it should not be forgotten either that issues such as cost, ease of manufacture, reliability and yield are often decisive factors in 'tipping the balance' in favour of one technology over another, even if 'performance' suffers as a result. Nowhere is this message more relevant than in the field of high speed digital logic where, precisely because of these considerations, the preeminence of Si integrated circuits in everyday use looks set to continue for as long as anyone cares to imagine.

References.

[1]. High Speed Electronic Devices, S.M. Sze (Ed.), John Wiley and Sons (1990)
[2]. Physics of Semiconductor devices (2nd Edition), S.M. Sze, John Wiley and Sons (1981)
[3]. H. Fukui, IEEE Trans. Electron. Dev., **26**, 1032 (1979)
[4]. The Handbook on Semiconductors (Volume 4, Device Physics), C. Hilsum (Ed.), North Holland (1981)
[5]. Applications of GaAs MESFETs, R. Soares, J. Graffeuil and J. Obregon (Eds.), Artech House (1983)
[6]. G. Jerinic, J. Fines, M. Cobb and M. Schindler, Int. J. of Infrared and Millimeter Waves, **6**, 79 (1985)
[7]. Optoelectronics: An Introduction, J. Wilson and J.W. Hawkes, Prentice Hall (1983)
[8]. R.J. McIntyre, IEEE Trans. Electron Dev., **13**, 164 (1966)
[9]. S. Adachi, J. Appl. Phys., **58**, R1 (1985)
[10]. T. Ishibashi and Y. Yamauchi, IEEE Trans. Electron Dev., **35**, 401 (1988)
[11]. R.N. Nottenberg, Y.K. Chen, M.B. Panish, D.A. Humphrey and R. Hamm, IEEE Electron Dev. Lett., **10**, 30 (1989)
[12]. T. Sugeta and T. Ishibashi, Semiconductors and Semimetals, **30**, 195 (1990)
[13]. R.W. Keyes, Physics Today (August issue), 42 (1992)
[14]. J.C. Bean, J. Electron. Mat., **19**, 1055 (1990)
[15]. R. People and J.C. Bean, Appl. Phys. Lett., **47**, 322 (1985)
[16]. G.L. Patton, S.S. Iyer, S.L. Delage, S. Tiwari and J.M.C. Stork, IEEE Electron Dev. Lett., **9**, 165 (1988)
[17]. H. Temkin, J.C. Bean, A. Antreasyan and R. Leibenguth, Appl. Phys. Lett., **52**, 1089 (1988)
[18]. H. Hirayama, M. Hiroi, K. Koyama and T. Tatsumi, Appl. Phys. Lett., **56**, 2645 (1990)
[19]. S.S. Iyer, G.L. Patton, D.L. Harame, J.M.C. Stork, E.F. Crabbe and B.S. Meyerson, Thin Solid Films, **184**, 153 (1990)
[20]. A. Pruijmboom, J.W. Slotboom, D.J. Gravesteijn, C.W. Fredriksz, A.A. van Gorkum, R.A. van de Heuval, J.M.L. van Rooij-Mulder, G. Streutker and G.F.A. van deWalle, IEEE Electron Dev. Lett., **12**, 357 (1991)
[21]. A. Gruhle, H. Kibbel, U. Konig, U. Erben and E. Kasper, IEEE Electron Dev. Lett., **13**, 206 (1992)
[22]. S. Hiyamizu, Semiconductors and Semimetals, **30**, 53 (1990)
[23]. T. Mimura, Semiconductors and Semimetals, **30**, 157 (1990)
[24]. R. Dingle, H.L. Stormer, A.C. Gossard and W. Weigman, Appl. Phys. Lett., **33**, 665 (1978)
[25]. L. Pfeiffer, K.W. West, H.L. Stormer and K.W. Baldwin, Appl. Phys. Lett., **55**, 1888 (1989)
[26]. P. Hendriks, E.A.E. Zwaal, J.E.M. Haverkort and J.H. Wolter, SPIE Physical Concepts of Materials for Novel Optoelectronic Device Applications, **1362**, 217 (1990)

[27]. P.H. Ladbrooke, GEC J. Res., **4**, 115 (1986)
[28]. K.H.G. Duh, M.W. Pospieszalski, W.F. Kopp, P. Ho, A.A. Jabra, P.C. Chao, P.M. Smith, L.F. Lester, J.M. Ballingall and S. Weinreb, IEEE Trans. Electron Dev., **35**, 249 (1988)
[29]. T. Nakanisi, Semiconductors and Semimetals, **30**, 105 (1990)
[30]. U.K. Mishra, A.S. Brown, S.E. Rosenbaum, C.E. Hooper, M.W. Peirce, M.J. Delaney, S. Vaughn and K. While, IEEE Electon Dev. Lett., **9**, 647 (1988)
[31]. U.K. Mishra, A.S. Brown, M.J. Delaney, P.T. Greiling and C.F Krumm, IEEE Trans. Micro. Theor. Tech., **37**, 1279 (1989)
[32]. K.L. Tan, D.C. Streit, P.D. Chow, R.M. Dia, A.C. Han, P.H. Liu, D. Garske and R. Lai, Proceedings of the 1991 IEEE International Electron Devices Meeting, 239 (1991)
[33]. T. Mishima, C.W. Fredriksz, G.F.A. Van de Walle, D.J. Gravesteijn, R.A. Van den Heuvel and A.A. Van Gorkum, Appl. Phys. Lett., **57**, 2567 (1990)
[34]. E. Murakami, K. Nakagawa, A. Nishida and M. Miyao, IEEE Electron Dev. Lett., **12**, 71 (1991)
[35]. F. Schaffler, D. Tobben, H-J. Herzog, G. Abstreiter and B. Hollander, Semicond. Sci. Technol., **7**, 260 (1992)
[36]. J.D. Cressle, J.H. Comfort, E.F. Crabbe, G.L. Patton, W. Lee, J.Y.C. Sun, J.M.C. Stork and B.S. Meyerson, IEEE Trans. Electron Dev. Lett., **12**, 166 (1991)
[37]. J.M. Woodcock, J.J. Harris and J.M. Shannon, Physica B, **134**, 111 (1985)
[38]. J.R. Hayes, A.F.J. Levi and W. Weigmann, Phys. Rev. Lett., **54**, 1570 (1985)
[39]. M. Heiblum, M.I. Nathan, D.C. Thomas and C.M. Konedler, Phys. Rev. Lett., **55**, 2200 (1985)
[40]. S. Muto, K. Imamura, N. Yokoyama, S. Hiyamizu and H. Nishi, Electron. Lett., **21**, 555 (1985)
[41]. P.H. Beton, A.P. Long and M.J. Kelly, J. Appl. Phys., **65**, 3076 (1989)
[42]. A.F.J. Levi and T.H. Chiu, Appl. Phys. Lett., **51**, 984 (1987)
[43]. P.M. Matthews, M.J. Kelly, V.J. Law, D.G. Hasko, M. Pepper, W.M. Stobbs, H. Ahmed, D.C. Peacock, J.E.F. Frost, D.A. Ritchie and G.A.C. Jones, Phys. Rev. B., **42**, 11415 (1990)
[44]. G.H. Dohler, IEEE J. Quantum Electron., **22**, 1682 (1986)
[45]. K. Yamaguchi, Y. Shiraki, Y. Katayama and Y. Murayama, Jpn. J. Appl. Phys. Suppl., **22-1**, 267 (1983)
[46]. E.F. Schubert, A. Fischer and K. Ploog, IEEE Trans. Electron. Dev., **33**, 625 (1986)
[47]. K. Nakagawa, A.A. Van Gorkum and Y. Shiraki, Appl. Phys. Lett., **54**, 1869 (1989)
[48]. P.D. Coleman, J. Freeman, H. Morkoc, K. Hess, B.G. Streetman and M. Keever, Appl. Phys. Lett., **40**, 493 (1982)
[49]. A. Kastalsky, J.H. Abeles, R. Bhat, W.K. Chan and M. Kosa, Appl. Phys. Lett., **48**, 71 (1986)
[50]. M.R. Hueschen, N. Moll and A. Fischer-Colbrie, Appl. Phys. Lett., **57**, 386 (1990)
[51]. S. Luryi, P.M. Mensz, M.R. Pinto, P.A. Garbinski and A.Y. Cho, Appl. Phys. Lett., **57**, 1787 (1990)

[52]. R.J. Malik, T.R. Aucoin, R.L. Ross, K. Board, C.E.C. Wood and L.F. Eastman, Electron. Lett, **16**, 836 (1980)
[53]. J.M. Shannon, Philips J. of Res., **40**, 239 (1985)
[54]. M.J. Kearney and I. Dale, GEC. J. of Res., **8**, 1 (1990)
[55]. N.R. Couch and M.J. Kearney, J. Appl. Phys., **66**, 5083 (1989)
[56]. Z. Greenwald, D.W. Woodward, A.R. Calawa and L.F. Eastman, Solid St. Electron., **31**, 1211 (1988)
[57]. N.R. Couch, H. Spooner, P.H. Beton, M.J. Kelly, M.E. Lee, P.K. Rees and T.M. Kerr, IEEE Electron Dev. Lett., **10**, 288 (1989)
[58]. N.R. Couch and H. Spooner, GEC. J. of Res., **7**, 34 (1989)
[59]. W. Behr and J.F. Luy, IEEE Electron. Dev. Lett., **11**, 206 (1990)
[60]. M.J. Kearney, N.R. Couch, J.S. Stephens and R.S. Smith, Electron Lett., **28**, 706 (1992)
[61]. J.I. Nishizawa, K. Motoya and Y. Okuno, IEEE Trans. Micro. Theor. Tech., **26**, 1029 (1978)
[62]. M.E. Elta and G.I. Haddad, IEEE Trans. Micro. Theor. Tech., **27**, 442 (1979)
[63]. M. Pobl, C. Dalle, J. Freyer and W. Harth, Electron. Lett., **26**, 1542 (1990)
[64]. C. Kidner, H. Eisele and G.I. Haddad, Electron Lett., **28**, 511 (1992)
[65]. M.E. Elta, H.R. Fetterman, W.V. Macropoulos and J.J. Lambert, IEEE Electron Dev. Lett., **1**, 115 (1980)
[66]. J.F. Luy, H. Jorke, H. Kibbel, A. Casel and E. Kasper, Electron. Lett., **24**, 1386 (1988)
[67]. W.T. Tsang, Semiconductors and Semimetals, **24**, 397 (1987)
[68]. N. Chand, E.E. Becker, J.P. Van der Ziel, S.N.G. Chu and N.K. Dutta, Appl. Phys. Lett., **58**, 1704 (1991)
[69]. H. Soda, K. Iga, C. Kitahara and Y. Suematsu, Jpn. J. Appl. Phys., **18**, 2329 (1979)
[70]. F. Koyama, S. Kinoshita and K. Iga, Appl. Phys. Lett., **55**, 221 (1989)
[71]. R.C. Miller, B. Schwartz, L.A. Koszi and W.R. Wagner, Appl. Phys. Lett., **33**, 721 (978)
[72]. J.E. Bowers and C.A. Burrus, J. Lightwave Tech., **5**, 1339 (1987)
[73]. S.R. Forrest, O.K. Kim and R.G. Smith, Appl. Phys. Lett., **41**, 95 (1982)
[74]. Y. Matsushima, A. Akiba, K. Sakai, Y. Kushiro, Y. Noda and K. Utaka, Electron. Lett., **18**, 945 (1982)
[75]. L.E. Tarof, D.G. Knight, K.E. Fox, C.J. Miner, N. Puetz and H.B. Kim, Appl. Phys. Lett., **57**, 670 (1990)
[76]. F. Capasso, W.T. Tsang, A.L. Hutchinson and G.F. Williams, Appl. Phys. Lett., **40**, 38 (1982)
[77]. N. Susa and H. Okamoto, Jpn. J. Appl. Phys., **23**, 317 (1984)
[78]. T. Kagawa, Y. Kawamura, H. Asai and M. Naganuma, Appl. Phys. Lett., **57**, 1895 (1990)
[79]. H. Okamoto, Jpn. J. Appl. Phys., **26**, 315 (1987)
[80]. B.F. Levine, C.G. Bethea, G. Hasnian, V.O. Shen, E. Pelve, R.R. Abbott and S.J. Hsieh, Appl. Phys. Lett., **56**, 851 (1990)
[81]. S.R. Andrews and B.A. Miller, J. Appl. Phys., **70**, 993 (1991)

[82]. D.S. Chemla, D.A.B. Miller and P.W. Smith, Semiconductors and Semimetals, **24**, 279 (1987)
[83]. D.A.B. Miller, D.S. Chemla, T.C. Damen, A.C. Gossard, W. Wiegmann, T.H. Wood and C.A. Burrus, Phys. Rev. B., **32**, 1043 (1985)
[84]. T.H. Wood, C.A. Burrus, D.A.B. Miller, D.S. Chemla, T.C. Damen, A.C. Gossard and W. Wiegmann, Appl. Phys. Lett., **44**, 16 (1984)
[85]. J.S. Weiner, D.A.B. Miller, D.S. Chemla, T.C. Damen, C.A. Burrus, T.H. Wood, A.C. Gossard and W. Weigmann, Appl. Phys. Lett., **47**, 1148 (1985)
[86]. T.H. Wood, C.A. Burrus, R.S. Tucker, J.S. Weiner, D.A.B. Miller, D.S. Chemla, T.C. Damen, A.C. Gossard and W. Weigmann, Electron. Lett., **21**, 693 (1985)
[87]. H. Nagai, M. Yamanishi, Y. Kan and I. Suemune, Electron. Lett., **22**, 888 (1986)
[88]. D.A.B. Miller, D.S. Chemla, T.C. Damen, A.C. Gossard, W. Weigmann, T.H. Wood and C.A. Burrus, Appl. Phys. Lett., **45**, 13 (1984)
[89]. D.A.B. Miller, J.E. Henry, A.C. Gossard and J.H. English, Appl. Phys. Lett., **49**, 821 (1986)
[90]. R. Tsu and L. Esaki, Appl. Phys. Lett., **22**, 562 (1973)
[91]. L.L. Chang, L. Esaki and R. Tsu, Appl. Phys. Lett., **24**, 593 (1974)
[92]. B. Ricco and M.Ya. Azbel, Phys. Rev. B., **29**, 1970 (1984)
[93]. C.I. Huang, M.J. Paulus, C.A. Bozada, S.C. Dudley, K.R. Evans, C.E. Stutz, R.L. Jones and M.E. Cheney, Appl. Phys. Lett, **51**, 121 (1987)
[94]. A.A. Lakhani, R.C. Potter, D. Beyea, H.H. Hier, E. Hempfling, L. Aina and J.M. O'Conner, Electron. Lett., **24**, 153 (1988)
[95]. T.C.L.G. Sollner, P.E. Tannenwald, D.D. Peck and W.D. Goodhue, Appl. Phys. Lett., **45**, 1319 (1984)
[96]. I. Song and D.S. Pan, IEEE Trans. Electron Dev., **35**, 2315 (1988)
[97]. V.P. Kesan, A. Mortazawi, D.R. Miller, V.K. Reddy, D.P. Neikirk and T. Itoh, IEEE Trans. Micro. Theor. Tech., **37**, 1933 (1989)
[98]. S. Javalagi, V. Reddy, K. Gullapalli and D. Neikirk, Electron. Lett., **28**, 1699 (1992)
[99]. E.R. Brown, C.D. Parker and T.C.L.G. Sollner, Appl. Phys. Lett., **54**, 934 (1989)
[100]. C. Kidner, I. Mehdi, J.R. East and G.I. Haddad, IEEE Trans. Micro. Theor. Tech., **38**, 864 (1990)
[101]. T.C.L.G. Sollner, W.D. Goodhue, P.E. Tannenwald, C.D. Parker and D.D. Peck, Appl. Phys. Lett., **43**, 588 (1983)
[102]. E.R. Brown, T.C.L.G. Sollner, W.D. Goodhue and C.D. Parker, Appl. Phys. Lett., **50**, 83 (1987)
[103]. E.R. Brown, T.C.L.G. Sollner, C.D. Parker, W.D. Goodhue and C.L. Chen, Appl. Phys. Lett., **55**, 1777 (1989)
[104]. E.R. Brown, J.R. Soderstrom, C.D. Parker, L.J. Mahoney, K.M. Molvar and T.C. McGill, Appl. Phys. Lett., **58**, 2291 (1991)
[105]. S.K. Diamond, E. Ozbay, M.J.W Rodwell, D.M. Brown, Y.C. Pao, E. Wolak and J.S. Harris, IEEE Electron Dev. Lett., **10**, 104 (1989)
[106]. A.A. Lakhani, R.C. Potter and H.S. Hier, Electron. Lett., **24**, 681 (1988)

[107]. F. Capasso, S. Sen., F. Beltram, L.M. Lunardi, A.S. Vengurlekar, P.R. Smith, N.J. Shah, R.J. Malik and A.Y. Cho, IEEE Trans. Electron. Dev., **36**, 2065 (1989)

[108]. T. Mori, T. Adachihara, M. Takatsu, H. Ohnishi, K. Imamura, S. Muto and N. Yokoyama, Electron Lett., **27**, 1523 (1991)

[109]. Quantum Semiconductor Structures, C. Weisbuch and B. Vinter, Academic (1991)

New Concepts to fabricate semiconductor quantum wire and quantum dot structures

Klaus H. Ploog* and Richard Nötzel**

* Paul-Drude-Institut für Festkörperelektronik
  O - 1086 Berlin, Federal Republic of Germany
** Max-Planck-Institut für Festkörperforschung
  W - 7000 Stuttgart 80, Federal Republic of Germany

1 Introduction

The lectures presented during this School have shown that the quantum confinement of electrons and holes in artificial low-dimensional semicondctor structures strongly modifies their electronic properties which has an important impact on the performance of high-speed electronic and photonic devices. The condition for the occurrence of new electronic phenomena in such structures is that the size of the active region being smaller than the coherence and elastic scattering lengths of the carriers. Advanced crystal growth techniques, such as molecular beam epitaxy (MBE) and metalorganic vapor phase epitaxy (MO VPE), are now routinely used to fabricate artificially layered semiconductors of high structural perfection which exhibit a (periodic) modulation in chemical composition perpendicular to the crystal surface down to atomic dimensions [1]. In these quasi-two-dimensional (2D) semiconductor structures the motion of free carriers is quantized within the layer plane. This quantum confinement of carriers results in a distinct modification of the energy dependence of their density of states g(E), i.e. the number of the allowed states per energy interval (Fig. 1). In bulk semiconductors, where the minimum of the characteristic lengths is much larger than the mean free path of electrons, the g(E) curve has a parabolic shape and it is zero at the band edge (minimum energy E_g). In different semiconductor materials only the curvature of the parabolic g(E) behavior changes. In contrast, the steplike g(E) behavior of 2D systems has a finite non-zero value of the density of states even at the minimum energy. In addition, the distinct shape of the steplike g(E) curve can be tailored by appropriate choice the constituent layer thicknesses. With further reduction of the dimensionality the g(E) distribution changes to

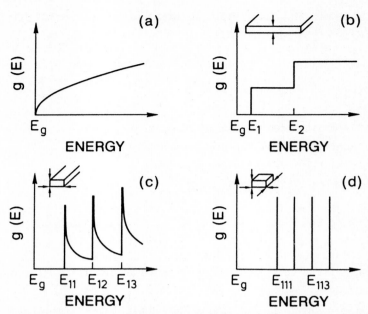

Fig. 1 Schematic illustration of the energy dependence of the density of states g(E) for (a) bulk (3D), (b) quantum well (2D), (c) quantum wire (1D), and quantum dot (0D) structures. The insets show rectangular quantum-confined configurations of the respective structures.

a sawtooth-behavior for quantum wires and to a singular behavior for quantum dots (Fig. 1). As a result, the artificial arrangement of the constituent atoms in these low-dimensional semiconductor structures leads to completely different shapes of the g(E) curves and hence to novel electronic properties even for structures made of the most familiar semiconductor materials. In addition to the drastic changes of the g(E) curves in these low-dimensional semiconductor structures, the wave nature of the electron (hole) becomes fundamental to the studied phenomena if the structural features are reduced to below 500 Å in size, i.e. to the range of the de Broglie wavelength where only a few size-defined modes are populated.

However, as yet the fabrication of quantum wire and quantum dot structures has mainly been tried by means of subsequent lateral patterning of 2D heterostructures and quantum wells with nanoscale lithographic techniques [3-5]. The minimum lateral dimensions achieved in such structures are much larger than the vertical ones given by epitaxy, hence leading to relatively small separations of the subband energies. These narrow subband spacings are then often masked by the level broadening due to irregularities and defects introduced during the patterning process. To reduce especially the defect density, several methods for direct fabrication of quantum-wire and quantum-dot structures based on epitaxial growth have been exploited. In these structures, lateral dimensions comparable to the vertical ones can be achieved. Hence, they allow in principle for large subband separations, which are required for optical and electrical device applications [6]. The cleaved-edge overgrowth which forms one-dimensional (1D) structures by growth on the cleaved edge of a 2D heterostructure [7], the overgrowth of prepatterned V-grooved substrates [8,9], and the generation of supersteps [10,11] have been investigated. The formation of lateral superlattices by growth on vicinal substrates has become most prominent. Here, the lateral potential is provided by depositing fractions of a monolayer of material with higher (AlAs) [12] or lower (InAs) [13] bandgap compared to GaAs which accumulates at the step edges on vicinal GaAs substrates inclined by a small misorientation to singular surfaces. Due to the height of these steps of only one monolayer, however, efficient lateral confinement requires the creation of lateral columns of material with alternating composition by subsequent deposition of regular fractional monolayers. This imposes tremendous demands on the stability of the growth rates and step uniformity which is hardly achievable due to a poor control of the local misorientation and the kink formation.

In this paper we describe a new concept to directly synthesize GaAs quantum wire and quantum dot structures in an AlAs matrix by conventional elemental-source MBE which overcomes the difficulties associated with the other direct-fabrication methods [14]. The concept is based on the evolution of well-ordered surface and interface structures on (111), (211), and (311) GaAs during epitaxy. Under typical MBE growth conditions, these nominally flat surfaces with high surface energy break up into regular arrays of macrosteps (facets) with spacings and heights of nanometer dimensions in order to lower the surface energy [15]. Selforganization during growth leads to a conservation of the surface corrugation during homoepitaxy and to a unique phase change during heteroepitaxy of AlAs on GaAs and vice versa. The resulting GaAs/AlAs multilayer structures then contain periodically alternating wide and narrow regions of GaAs and AlAs which form symmetric and asymmetric quantum-dot structures on (111) and (211) substrates and quantum-wire structures on (311) substrates. The observed optical properties verify the additional lateral size quantization in these GaAs/AlAs heterostructures.

2 Fabrication and Structural Properties

After the standard process of oxide removal from the GaAs sustrate at 580-590°C in the MBE growth chamber the originally flat (111), (211), and (311) surfaces break up into ordered arrays of macrosteps. This distinct faceting which can be monitored directly by reflection high-energy electron diffraction (RHEED) (Fig. 3) is preserved during epitaxial growth under standard conditions, i.e., growth rate 0.3-1.2 μm/hr, group-V-to-III flux ratio 5:1, and substrate temperature 580-600°C.

In order to facilitate the discussion of the RHEED observations, we briefly specify the reciprocal space of ordered surface structures, which is directly imaged by the RHEED patterns [16,17]. The asymmetric step array of upward or downward steps is presented in Fig. 2a. The lateral periodicity l and the step height h are deduced from the horizontal splitting of the streaks into slashes and from the length of the streaks normal to the surface, respectively. The lattice constant a of the terrace plane is determined from the separation of the corresponding streaks which are oriented along the lines connecting the intensity maxima of the slashes (dashed lines). The terrace width w is estimated from the full-width at half-maximum of the streaks corresponding to the

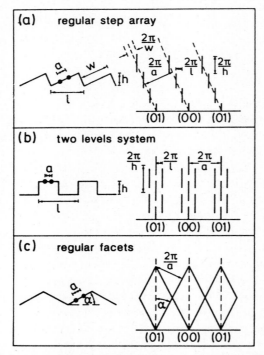

Fig. 2 Periodic surface structures with their representation in reciprocal space (a) regular step array, (b) two-level system, and (c) regular facets.

terrace planes. The reciprocal space of a two-level system with symmetrically arranged upward and downward steps (Fig. 2b) exhibits streaks which are split alternatingly into satellites and along their length. Intensity maxima of the satellites correspond to intensity minima of the integral order streaks and vice versa. The lateral periodicity l and step height h are deduced from the separation of the satellites and the splitting along the streaks, respectively. The reciprocal space of regular facets exhibits a periodic arrangement of tilted streaks (Fig. 2c). The tilt angle of the facet plane with respect to the nominal surface is determined from the tilt angle of the streak to the surface normal. The lattice constant a of the facet plane is determined from the streak separation.

The RHEED pattern of the (211) surface shows a reversible facetting of the flat surface at temperatures above 590°C (the transition occurs continuously in the range 550-590°C). Observation along the [01$\bar{1}$] azimuth above 590°C (Fig. 3a) exhibits a stepped surface with the step edges along [01$\bar{1}$]. The lateral periodicity l and step height h of the regular step array amount to l = 9.8 Å (= $2a_{111}$) and h = 2.3 Å. The ter-

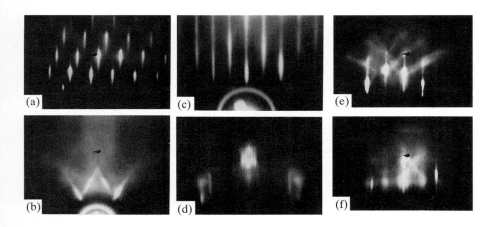

Fig. 3 Reflection high-energy electron (RHEED) patterns taken at 30 keV of the (211) GaAs surface (a) along [01$\bar{1}$], (b) along [$\bar{1}$11], of the (311) GaAs surface (c) along [01$\bar{1}$], (d) along [$\bar{2}$33], and of the (111) GaAs surface (e) along [1$\bar{1}$0], and (f) along [11$\bar{2}$].

race width w is estimated to w = 9.6Å. The (111) surface configuration in [01$\bar{1}$] projection of the terrace plane follows from the lattice constant a = 3.4 Å (= a_{211}). Below 550°C the RHEED pattern taken along [01$\bar{1}$] becomes diffuse showing a high degree of disorder of the flat (211) surface which is also revealed by the broad 00 streak for the perpendicular [$\bar{1}$11] azimuth. Above 590°C the RHEED pattern taken along [$\bar{1}$11] (Fig. 3b) shows again slashes tilted by 30° to the surface normal, indicating the presence of {110} facets. The extension of these slashes is comparable to those observed along [01$\bar{1}$] (Fig. 3a) thus evidencing the commensurability of the steps and facets. As a consequence, the stepped (211) surface comprises two sets of {110} facets along [$\bar{1}$11] and alternating (111) terrace planes of 6.9 Å width (= $2a_{211}$) and (001) steps of 4 Å width (= a_{100}) along [01$\bar{1}$] forming asymmetric pyramids of 2.3 Å height (= $2d_{221}$).

The RHEED pattern of the (311) surface taken along [01$\bar{1}$] reveals a pronounced streaking (Fig. 3c) which indicates a high density of steps along the perpendicular [$\bar{2}$33] direction. Taking the [$\bar{2}$33] azimuth parallel to the steps (Fig. 3d) the RHEED pattern directly images the reciprocal lattice of an almost perfect two level system oriented along [$\bar{2}$33]. Contrary to the case of the (211) surface, the RHEED pattern of the stepped (311) surface is stable down to room temperature. The lateral periodicity l and step height h amount to l = 32 Å (= $8a_{110}$) and h = 10 Å. The extinction of the main streak intensity for maximum intensity of the satellites evidences the high degree of ordering. These experimental parameters agree perfectly with our model describing the surface to be composed of (311) terraces of 4 Å width (= a_{110}) and two sets of (33$\bar{1}$) and ($\bar{3}$1$\bar{3}$) facets corresponding to upward and downward steps of 10.2 Å height (= $6d_{311}$) having a low surface energy (see Fig. 5a) [18].

The RHEED pattern of the (111) surface shows a pronounced splitting along the integral order streaks which is observed along arbitrary azimuthal directions (Figs. 3e,f for the [1$\bar{1}$0] and [11$\bar{2}$] directions). Hence, this splitting reflects the height of the surface structure of the (111) plane amounting to 13 Å. Splitting across the streaks is not resolved, indicating the lateral periodicity exceeding the transfer width of our RHEED system of about 100 Å. From the high symmetry of the [111] crystallographic direction the (111) surface is assumed to break up into symmetric pyramids of 13 Å height (= $4d_{111}$).

As the surface and interface ordering on (311) GaAs is at present best understood [19], we will discuss this orientation in some detail. During growth of (311) GaAs/AlAs multilayer structures the RHEED intensity dynamics show a pronounced oscillation at the onset of GaAs and AlAs growth, respectively (Fig. 4). This oscillation corresponds to the deposition of three (311) monolayers (ML), i.e., lattice planes, as deduced from the growth rates. During deposition of the next 3 ML the intensity approaches the value found in the RHEED pattern of the stable stepped surface during growth. The whole sequence then comprises the deposition of six (311) lattice planes, i.e., 10.2 Å, as shown in the inset of Fig. 4 for different growth rates of GaAs and AlAs. This sequence results from a phase change of the surface corrugation during the heterogeneous deposition of GaAs on AlAs and vice versa. The phase change includes quasi-filling of the corrugation during the first 3 ML deposition and

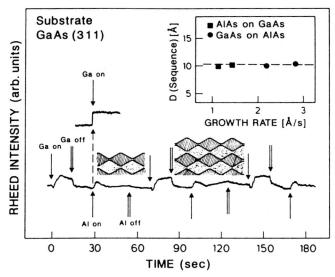

Fig. 4 Reflection high-energy electron diffraction (RHEED) intensity dynamics taken along [233] during growth of (311) GaAs/AlAs multilayer structures. The upper curve shows the RHEED intensity during deposition of GaAs on GaAs. The inset shows the deposited layer thickness for different growth rates of GaAs on AlAs and vice versa until stable growth conditions are reached. The schematic drawing images the GaAs/AlAs multilayer structure resulting from the phase change of the surface corrugation during the heterogeneous deposition of the first monolayers of GaAs and AlAs on GaAs.

re-arrangement of the stepped surface during the next 3 ML deposition. After completion of the phase change, the growth continues layer by layer with conservation of the surface corrugation as indicated by the constant RHEED intensity of the stepped surface during homogenous growth and by the abrupt change of the RHEED intensity from the non-growing GaAs (AlAs) surface to the growing GaAs (AlAs) surface within the deposition of 1 ML (see upper curve of Fig. 4). We assume that this phase change is induced by strain which makes the heterogenous growth on the facets energetically more favorable. The completed GaAs/AlAs multilayer structure then consists of well ordered alternating thicker and thinner regions of GaAs and AlAs oriented along [$\bar{2}$33] (Fig. 4). This unique arrangement indeed forms an as-grown GaAs quantum-wire structure in an AlAs matrix (see Fig. 5). A very similar behavior of the RHEED intensity is observed for the growth of GaAs/AlAs multilayer structures on (111) and (211) oriented substrates, so that the phase change of the respective surface corrugation is effective also for these orientations.

The existence of GaAs quantum-wire structures in the AlAs matrix is confirmed by high-resolution transmission-electron microscopy (HREM) [15],

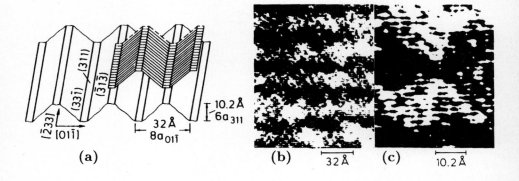

Fig. 5 (a) Schematic illustration of the stepped GaAs (311) surface. The upper surface (shaded) illustrates the phase change of the surface corrugation during heterogeneous growth of GaAs on AlAs and vice versa. (b), (c) High-resolution transmission-electron microscope (HREM) image of a 15Å/13Å (311) GaAs/AlAs multilayer structure viewed along [$\bar{2}$33].

as shown in Fig. 5b,c. Because of the unintentional misorientation of the (311) substrates by 1° and the microroughness of the interfaces the contrast between GaAs and AlAs is not too sharp. Nevertheless, detailed X-ray diffraction measurements reveal that the structural perfection of GaAs/AlAs multilayer structures grown on (311), (211), and (111) substrates is as excellent as that of (100) reference samples [19]. In addition, the average GaAs and AlAs layer thickness (neglecting the specific interface corrugation) are found to be comparable to that of the reference sample, i.e., the overall sticking coefficient of Ga an Al is unity for the (311), (211), (111), and (100) orientation.

3 Electronic Properties

The photoluminescence (PL) lines of the GaAs quantum-wire and quantum-dot structures are always shifted to lower energies as compared to a (100) multiple quantum-well (MQW) reference sample grown side by side with the (311), (211), and (111) samples. This redshift arises from the fact that the luminescence originates from the respective wider wire and dot regions. The redshift of the PL obtained from (311), (211), and (111) GaAs/AlAs multilayer structures with the same average GaAs and AlAs thicknesses increases with the height of the surface corrugation in the different orientations (see Fig. 6 and Table 1 for 46Å GaAs multilayer structures). In addition, we observe an enhancement of the light-hole (lh) exciton continuum energies with the height of the surface corrugation (Table 1). The lh exciton continuum energies are deduced from the energy spacing between the lh-exciton transition and the supplementary lh transitions observed in the photoluminescence excitation (PLE)

Tab. 1 Dependence of the luminescence redshift and light-hole (lh) exciton continuum energy on the height of the surface corrugations for various orientations.

Orientation	(100)	(211)	(311)	(111)
Height of surface corrugation [Å]	0	2.3	10.2	13.1
Red-shift of luminescence [meV]	0	16	24	38
lh-exciton continuum energy [meV]	15	27	29	36

spectra (Fig. 6). Their substantial increase in the (211), (311), and (111) structures directly reflects the additional lateral confinement [20] provided by the specific interface corrugation. The lh character of these transitions is evidenced by the observed negative degree of circular polarization which excludes their origin from "forbidden" transitions. The increased stability of the confined excitons is also manifested by the strong exciton-phonon interaction which is revealed by the observation of hot-exciton relaxation [21]. When the detection wavelength is set to the high-energy side of the PL line, strong LO and TA phonon related lines are resolved in the PLE spectra of the (311), (211), and (111) samples (Fig. 6), because of the high probability of the laterally confined excitons created above the bandgap to relax as a whole. The energy threshold for damping of the phonon lines can be used to estimate the 1D confinement energy for excitons. The striking result of this estimate is that the 1D exciton confinement energy reaches values up to 90 meV for 43 Å GaAs (311) quantum-wire structures.

The PLE spectra of the (211) and (311) samples (Fig. 6b,c) exhibit a pronounced polarization anisotropy of the excitonic resonances. The lh resonance is more pronounced in the spectrum taken with the light polarized parallel to the [011] direction, whereas the heavy-hole (hh) resonance is more pronounced for the respective perpendicular polarization in agreement with the asymmetric interface corrugation in these structures [22]. The corresponding polarization behavior is observed also in PL. No optical anisotropy is observed for (100) and (111) samples as expected for quantum wells and also for the symmetric surface structure found for the (111) orientation. The additional lateral confinement existing in the (111) sample manifests itself in the strongly increased exciton continuum energy and in the appearance of phonon related lines in the PLE spectra. The lh exciton continuum energy is observed also in the

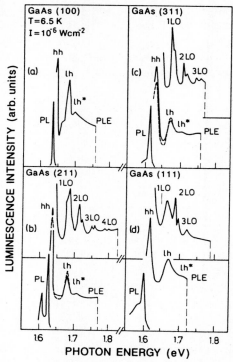

Fig. 6 Low-temperature photoluminescence (PL) and photoluminescence excitation (PLE) spectra of (a) (100), (b) (211), (c) (311), and (d) (111) 46Å/41Å GaAs/AlAs multilayer structures. In (b) and (c) the dashed lines show the PLE spectra for light polarized parallel to the [011] azimuth (parallel to the direction of lateral quantization) and the solid lines corres-pond to the respective perpendicular directions. The sharp LO- and TA-phonon related lines are resolved when the detection energy is set to the high energy side of the PL line.

absorption spectrum shown in Fig. 7, which is obtained by detecting the PL of the GaAs buffer layer. The enhanced hh resonance for (111) samples can be attributed to a peaked density of states given by the highly symmetric surface structure, thus most probably indicating the existence of quantum dots. This result is supported by the observed strong absorption minima at energies above the hh state and between the lh- and hh (n = 2) state, and by the absence of the 1 LO-phonon line in the PLE spectra (Fig. 6d), which coincides in energy with the minimum in absorption above the hh state (Fig. 7).

Finally, it is important to note that the GaAs quantum-wire and quantum-dot structures exhibit an extremely high luminescence intensity. In Fig. 8 we show that at 300K the integrated PL intensity of the (311) quantum-

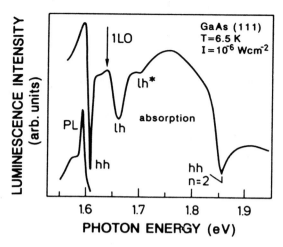

Fig. 7 Photoluminescence (PL) and absorption spectra of a 46Å/41Å GaAs (111) quantum-dot structure.

wire structure is by more than one order of magnitude larger than that of the reference (100) MQW sample. This behavior which does not degrade up to 400K arises from the additional lateral confinement and strong localization of the excitons in the quantum-wire structures. The nonradiative interface recombination is strongly suppressed due to the reduced spreading of the photogenerated carriers whose motion is now free only along the wire direction. This finding is important for the design of light emitting devices of high efficiency.

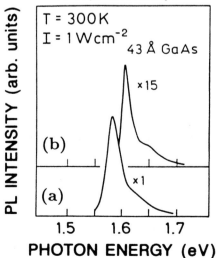

Fig. 8 Room-temperature luminescence of (a) a 43Å/47Å GaAs/AlAs (311) quantum-wire structure and (b) a 43Å/47Å GaAs/AlAs (100) multiple quantum-well structure.

4 Conclusion

The evolution of ordered surface and interface structures on (111), (211), and (311) oriented GaAs during molecular beam epitaxy offers the unique possibility to directly fabricate GaAs quantum-wire and quantum-dot arrays in an AlAs matrix. The formation of regular arrays of macro-steps with spacings and heights in the nanometer range is directly revealed by RHEED. Asymmetric pyramids are formed on the (211) surface, periodic channels on the (311) surface and symmetric pyramids on the (111) surface. The RHEED intensity dynamics exhibit a pronounced oscillation at the onset of GaAs and AlAs growth, respectively, due to a phase change of the surface corrugation during the heterogeneous deposition of AlAs on GaAs and vice versa. We assume that this phase change is induced by strain which plays an important role here and makes the heterogeneous growth on the facets energetically less favorable, so that the growth starts on the low-level terrace. As a consequence the complete structures consist of well ordered alternating thicker and thinner regions of GaAs and AlAs thus forming symmetric and asymmetric quantum-dot structures on (111) and (211) GaAs surfaces and quantum-wire structures on (311) GaAs surfaces. The observed redshift of the luminescence of GaAs/AlAs multilayer structures with the same average barrier and well widths as the (100) reference structures grown side by side correlates with the height of the surface corrugation of the different orientations reflecting the fact that the luminescence originates from transitions in the respective thicker GaAs regions. For a chosen orientation the redshift increases with decreasing well width accompanied by a strong enhancement of the integrated luminescence intensity compared to the (100) reference structures exceeding one order of magnitude for a 43Å GaAs (311) quantum-wire structure. This behavior shows the increasing influence of the interface corrugation for decreasing well width which results in a suppression of the nonradiative interface recombination due to the lateral localization of excitons. The observed enhancement of the exciton continuum energies which also correlates with the height of the surface corrugation of the different orientations directly indicates additional confinement due to the lateral potential introduced by the interface corrugation. The lateral confinement is further evidenced by the observation of hot-exciton relaxation, a pronounced optical anisotropy for the (211) and (311) samples, and by the enhanced heavy-hole resonance for (111) samples due to the peaked density states.

Although the results presented here are quite encouraging, the reproducible fabrication of low-dimensional semiconductors having quasi-1D and quasi-0D electronic properties remains one of the major challenges in spatially resolved materials synthesis and, moreover, in the entire field of microstructure materials science. Methods must be developed which enable the removal of materials atom-by-atom in well-defined spatial and geometrical arrangements without causing damage to the crystal surface. The techniques described here are only a first step towards the best choice for the fabrication of quantum wire and quantum dot structures. The search for methods and techniques to manipulate the atoms in a crystal one-by-one (growth, removal, displacement) should also be extended to other solid materials. Other promising approaches for the direct synthesis of quantum wires and quantum dots include semiconductor microcrystals grown in dielectric media [23], the selective coordinating epitaxy of mixed-valence metal compounds [24], the matrix isolation of clusters [25], the positioning of single atoms or molecules with a scanning tunneling microscope [26], and the host/guest chemistry, as for example binary semiconductor clusters in cages of a zeolite [27]. In the latter method the host can even provide a three-dimensional periodicity to form a "supra-molecular" composition and hence the overall quantum lattice.

Acknowledgement

This work was sponsored in part by the Bundesministerium für Forschung und Technologie of the Federal Republic of Germany.

References

[1] K.H. Ploog, this volume
[2] For a review on molecular beam epitaxy see: K. Ploog, Angew. Chem. Int. Ed. Engl. 27, 593 (1988)
[3] M.A. Reed, J.N. Randall, R.J. Aggarwal, R.J. Matyi, T.M. Moore, and A.E. Wetsel, Phys. Rev. Lett. 60, 535 (1988)
[4] K. Kash, B.P. Van der Gaag, D.D. Mahoney, A.S. Gozdz, L.T. Florez, J.P. Harbison, and M.D. Sturge, Phys. Rev. Lett. 67, 1326 (1991)
[5] M. Kohl, D. Heitmann, P. Grambow, and K. Ploog, Phys. Rev. Lett. 63, 2124 (1989)
[6] H. Sakaki, in: Localization and Confinement of Electrons in Semiconductors, edited by F. Kuchar, H. Heinrich, and G. Bauer, Springer Series in Solid-State Sciences, Vol. 97, (Springer Verlag, Heidelberg, 1990), p.2 (1990)
[7] D. Gershoni, J.S. Weiner, S.N.G. Chu, G.A. Baraff, J.M. Vandenberg, L.N. Pfeiffer, K. West, R.A. Logan, and T. Tanbun-Ek, Phys. Rev. Lett. 65, 1631 (1990)
[8] E. Kapon, D. Hwang, and R. Bhat. Phys. Rev. Lett. 63, 430 (1989)

[9] X.Q. Shen, M. Tanaka, and T. Nishinaga, (Workbook 7th Int. Conf. Molecular Beam Epitaxy, Aug. 24-28, Schwäbisch-Gmünd, FRG, 1992), Th P.8 (1992)
[10] M. Sato, K. Maehashi, H. Asahi, S. Hasegawa, and H. Nakashima, Superlattices and Microstructures 7, 279 (1990)
[11] T. Yamamoto, M. Inai, T. Takebe, and K. Kobayashi (Woorkbook 7th Int. Conf. Molecular Beam Epitaxy, Ref. 9), Th2.3 (1992)
[12] M. Sundaram, S.A. Chalmers. P.F. Hopkins, and A.C. Gossard, Science 254, 1326 (1991)
[13] O. Brandt, L. Tapfer, K. Ploog, R. Bierwolf, M. Hohenstein, F. Phillipp, H. Lage, and A. Heberle, Phys. Rev. B44, 8043 (1991)
[14] E. Corcoran, Sci. Am. 263, 74 (1990)
[15] R. Nötzel, N.N. Ledentsov, L. Däweritz, M. Hohenstein, and K. Ploog, Phys. Rev. Lett. 67, 3812 (1991)
[16] M. Henzler, Appl. Surf. Sci. 12, 450 (1982)
[17] M.G. Lagally, D.E. Savage, and M.C. Tringides, in Reflection High-Energy Electron Diffraction and Reflecting Electron Imaging of Surfaces, edited by P.K. Larsen and P.J. Dobson, NATO Advanced Study Institutes, Ser. B (Plenum, New York, 1988), p. 139 (1988)
[18] D.J. Chadi, Phys. Rev. B29, 785 (1984)
[19] R. Nötzel, N.N. Ledentsov, L. Däweritz, K. Ploog, and M. Hohenstein, Phys. Rev. B45, 3507 (1992)
[20] J.W. Brown and H.N. Spector, Phys. Rev. B35, 3009 (1987)
[21] S. Permogorov. Phys. Status Solidi B68, 9 (1975)
[22] D.S. Citrin and Y.C. Chang, Phys. Rev. B43, 11703 (1991)
[23] N. Chestnoy, R. Hull, and L.E. Brus, J. Chem. Phys. 85, 2237 (1986); A.I. Ekimov, I.A. Kudryavtsev, M.G. Ivanov, and A.L. Efros, J. Lumin. 46, 83 (1990)
[24] K. Takahashi, H. Tanino, and T. Yao, Jpn. J. Appl. Phys. 26, L97 (1987)
[25] T.P. Martin, Angew. Chem. Int. Ed. Engl. 25, 197 (1986)
[26] D.M. Eigler and E.K. Schweitzer, Nature 344, 524 (1990)
[27] G.D. Stucky and J.E. MacDougall, Science 247, 669 (1990)

Springer-Verlag and the Environment

We at Springer-Verlag firmly believe that an international science publisher has a special obligation to the environment, and our corporate policies consistently reflect this conviction.

We also expect our business partners – paper mills, printers, packaging manufacturers, etc. – to commit themselves to using environmentally friendly materials and production processes.

The paper in this book is made from low- or no-chlorine pulp and is acid free, in conformance with international standards for paper permanency.

Lecture Notes in Physics

For information about Vols. 1–379
please contact your bookseller or Springer-Verlag

Vol. 380: I. Tuominen, D. Moss, G. Rüdiger (Eds.), The Sun and Cool Stars: activity, magnetism, dynamos. Proceedings, 1990. X, 530 pages. 1991.

Vol. 381: J. Casas-Vazquez, D. Jou (Eds.), Rheological Modelling: Thermodynamical and Statistical Approaches. Proceedings, 1990. VII, 378 pages. 1991.

Vol. 382: V.V. Dodonov, V. I. Man'ko (Eds.), Group Theoretical Methods in Physics. Proceedings, 1990. XVII, 601 pages. 1991.

Vol. 384: M. D. Smooke (Ed.), Reduced Kinetic Mechanisms and Asymptotic Approximations for Methane-Air Flames. V, 245 pages. 1991.

Vol. 385: A. Treves, G. C. Perola, L. Stella (Eds.), Iron Line Diagnostics in X-Ray Sources. Proceedings, Como, Italy 1990. IX, 312 pages. 1991.

Vol. 386: G. Pétré, A. Sanfeld (Eds.), Capillarity Today. Proceedings, Belgium 1990. XI, 384 pages. 1991.

Vol. 387: Y. Uchida, R. C. Canfield, T. Watanabe, E. Hiei (Eds.), Flare Physics in Solar Activity Maximum 22. Proceedings, 1990. X, 360 pages. 1991.

Vol. 388: D. Gough, J. Toomre (Eds.), Challenges to Theories of the Structure of Moderate-Mass Stars. Proceedings, 1990. VII, 414 pages. 1991.

Vol. 389: J. C. Miller, R. F. Haglund (Eds.), Laser Ablation-Mechanisms and Applications. Proceedings. IX, 362 pages, 1991.

Vol. 390: J. Heidmann, M. J. Klein (Eds.), Bioastronomy - The Search for Extraterrestrial Life. Proceedings, 1990. XVII, 413 pages. 1991.

Vol. 391: A. Zdziarski, M. Sikora (Eds.), Ralativistic Hadrons in Cosmic Compact Objects. Proceedings, 1990. XII, 182 pages. 1991.

Vol. 392: J.-D. Fournier, P.-L. Sulem (Eds.), Large-Scale Structures in Nonlinear Physics. Proceedings. VIII, 353 pages. 1991.

Vol. 393: M. Remoissenet, M.Peyrard (Eds.), Nonlinear Coherent Structures in Physics and Biology. Proceedings. XII, 398 pages. 1991.

Vol. 394: M. R. J. Hoch, R. H. Lemmer (Eds.), Low Temperature Physics. Proceedings. X,374 pages. 1991.

Vol. 395: H. E. Trease, M. J. Fritts, W. P. Crowley (Eds.), Advances in the Free-Lagrange Method. Proceedings, 1990. XI, 327 pages. 1991.

Vol. 396: H. Mitter, H. Gausterer (Eds.), Recent Aspects of Quantum Fields. Proceedings. XIII, 332 pages. 1991.

Vol. 398: T. M. M. Verheggen (Ed.), Numerical Methods for the Simulation of Multi-Phase and Complex Flow. Proceedings, 1990. VI, 153 pages. 1992.

Vol. 399: Z. Švestka, B. V. Jackson, M. E. Machedo (Eds.), Eruptive Solar Flares. Proceedings, 1991. XIV, 409 pages. 1992.

Vol. 400: M. Dienes, M. Month, S. Turner (Eds.), Frontiers of Particle Beams: Intensity Limitations. Proceedings, 1990. IX, 610 pages. 1992.

Vol. 401: U. Heber, C. S. Jeffery (Eds.), The Atmospheres of Early-Type Stars. Proceedings, 1991. XIX, 450 pages. 1992.

Vol. 402: L. Boi, D. Flament, J.-M. Salanskis (Eds.), 1830-1930: A Century of Geometry. VIII, 304 pages. 1992.

Vol. 403: E. Balslev (Ed.), Schrödinger Operators. Proceedings, 1991. VIII, 264 pages. 1992.

Vol. 404: R. Schmidt, H. O. Lutz, R. Dreizler (Eds.), Nuclear Physics Concepts in the Study of Atomic Cluster Physics. Proceedings, 1991. XVIII, 363 pages. 1992.

Vol. 405: W. Hollik, R. Rückl, J. Wess (Eds.), Phenomenological Aspects of Supersymmetry. VII, 329 pages. 1992.

Vol. 406: R. Kayser, T. Schramm, L. Nieser (Eds.), Gravitational Lenses. Proceedings, 1991. XXII, 399 pages. 1992.

Vol. 407: P. L. Smith, W. L. Wiese (Eds.), Atomic and Molecular Data for Space Astronomy. VII, 158 pages. 1992.

Vol. 408: V. J. Martínez, M. Portilla, D. Sàez (Eds.), New Insights into the Universe. Proceedings, 1991. XI, 298 pages. 1992.

Vol. 409: H. Gausterer, C. B. Lang (Eds.), Computational Methods in Field Theory. Proceedings, 1992. XII, 274 pages. 1992.

Vol. 410: J. Ehlers, G. Schäfer (Eds.), Relativistic Gravity Research. Proceedings, VIII, 409 pages. 1992.

Vol. 411: W. Dieter Heiss (Ed.), Chaos and Quantum Chaos. Proceedings, XIV, 330 pages. 1992.

Vol. 412: A. W. Clegg, G. E. Nedoluha (Eds.), Astrophysical Masers. Proceedings, 1992. XX, 480 pages. 1993.

Vol. 413: Aa. Sandqvist, T. P. Ray (Eds.); Central Activity in Galaxies. From Observational Data to Astrophysical Diagnostics. XIII, 235 pages. 1993.

Vol. 414: M. Napolitano, F. Sabetta (Eds.), Thirteenth International Conference on Numerical Methods in Fluid Dynamics. Proceedings, 1992. XIV, 541 pages. 1993.

Vol. 415: L. Garrido (Ed.), Complex Fluids. Proceedings, 1992. XIII, 413 pages. 1993.

Vol. 416: B. Baschek, G. Klare, J. Lequeux (Eds.), New Aspects of Magellanic Cloud Research. Proceedings, 1992. XIII, 494 pages. 1993.

Vol. 417: K. Goeke P. Kroll, H.-R. Petry (Eds.), Quark Cluster Dynamics. Proceedings, 1992. XI, 297 pages. 1993.

Vol. 418: J. van Paradijs, H. M. Maitzen (Eds.), Galactic High-Energy Astrophysics. XIII, 293 pages. 1993.

Vol. 419: K. H. Ploog, L. Tapfer (Eds.), Physics and Technology of Semiconductor Quantum Devices. Proceedings, 1992. VIII, 212 pages. 1993.

New Series m: Monographs

Vol. m 1: H. Hora, Plasmas at High Temperature and Density. VIII, 442 pages. 1991.

Vol. m 2: P. Busch, P. J. Lahti, P. Mittelstaedt, The Quantum Theory of Measurement. XIII, 165 pages. 1991.

Vol. m 3: A. Heck, J. M. Perdang (Eds.), Applying Fractals in Astronomy. IX, 210 pages. 1991.

Vol. m 4: R. K. Zeytounian, Mécanique des fluides fondamentale. XV, 615 pages, 1991.

Vol. m 5: R. K. Zeytounian, Meteorological Fluid Dynamics. XI, 346 pages. 1991.

Vol. m 6: N. M. J. Woodhouse, Special Relativity. VIII, 86 pages. 1992.

Vol. m 7: G. Morandi, The Role of Topology in Classical and Quantum Physics. XIII, 239 pages. 1992.

Vol. m 8: D. Funaro, Polynomial Approximation of Differential Equations. X, 305 pages. 1992.

Vol. m 9: M. Namiki, Stochastic Quantization. X, 217 pages. 1992.

Vol. m 10: J. Hoppe, Lectures on Integrable Systems. VII, 111 pages. 1992.

Vol. m 11: A. D. Yaghjian, Relativistic Dynamics of a Charged Sphere. XII, 115 pages. 1992.

Vol. m 12: G. Esposito, Quantum Gravity, Quantum Cosmology and Lorentzian Geometries. XVI, 326 pages. 1992.

Vol. m 13: M. Klein, A. Knauf, Classical Planar Scattering by Coulombic Potentials. V, 142 pages. 1992.

Vol. m 14: A. Lerda, Anyons. XI, 138 pages. 1992.

Vol. m 15: N. Peters, B. Rogg (Eds.), Reduced Kinetic Mechanisms for Applications in Combustion Systems. X, 360 pages. 1993.

Vol. m 16: P. Christe, M. Henkel, Introduction to Conformal Invariance and Its Applications to Critical Phenomena. XV, 260 pages. 1993.

Vol. m 17: M. Schoen, Computer Simulation of Condensed Phases in Complex Geometries. X, 136 pages. 1993.

Printing: Druckerei Zechner, Speyer
Binding: Buchbinderei Schäffer, Grünstadt